LECTURES ON THE CALCULUS OF VARIATIONS

BY

OSKAR BOLZA

PROFESSOR OF MATHEMATICS
UNIVERSITY OF FREIBURG IM BREISGAU

THIRD EDITION

CHELSEA PUBLISHING COMPANY
NEW YORK, N. Y.

The present, third edition is a reprint of the first and second editions, in which the addenda have been incorporated into the text, in which certain additions and corrections made in the second edition have been retained, and in which the references to standard works have been updated to current (1973) editions It is published at New York, N. Y., 1973, and is printed on special 'long-Life' acid-free paper.

CIP

Library of Congress Cataloging in Publication Data

Bolza, Oskar, 1857-1942.

Lectures on the calculus of variations.

1. Calculus of variations. I. Title.

QA315.B7 1973 515'.64 73-16324

ISBN 0-8284-1145-X

First edition, Chicago, 1904. second edition, New York, 1961.

Printed in the United States of America

PREFACE

THE principal steps in the progress of the Calculus of Variations during the last thirty years may be characterized as follows:

1. A critical revision of the foundations and demonstrations of the older theory of the first and second variation according to the modern requirements of rigor, by WEIERSTRASS, ERDMANN, DU BOIS-REYMOND, SCHEEFFER, SCHWARZ, and others. The result of this revision was: a sharper formulation of the problems, rigorous proofs for the first three necessary conditions, and a rigorous proof of the sufficiency of these conditions for what is now called a "weak" extremum.

2. WEIERSTRASS'S extension of the theory of the first and second variation to the case where the curves under consideration are given in parameter-representation. This was an advance of great importance for all geometrical applications of the Calculus of Variations; for the older method implied —for geometrical problems—a rather artificial restriction.

3. WEIERSTRASS'S discovery of the fourth necessary condition and his sufficiency proof for a so-called "strong" extremum, which gave for the first time a complete solution, at least for the simplest type of problems, by means of an entirely new method based upon what is now known as "WEIERSTRASS'S construction."

These discoveries mark a turning-point in the history of the Calculus of Variations. Unfortunately they were given by WEIERSTRASS only in his lectures, and thus became known only very slowly to the general mathematical public. Chiefly under the influence of WEIERSTRASS'S theory a vigorous activity in the Calculus of Variations has set in during the last few years, which has led—apart from extensions and simplifications of WEIERSTRASS's theory—to the following two essentially new developments:

4. KNESER's theory, which is based upon an extension of certain theorems on geodesics to extremals in general. This new method furnishes likewise a complete system of sufficient conditions and goes beyond WEIERSTRASS's theory, inasmuch as it covers also the case of variable end points.

5. HILBERT's *a priori* existence proof for an extremum of a definite integral—a discovery of far-reaching importance, not only for the Calculus of Variations, but also for the theory of differential equations and the theory of functions.

To give a detailed account of this development was the object of a series of lectures which I delivered at the Colloquium held in connection with the summer meeting of the American Mathematical Society at Ithaca, N. Y., in August, 1901. And the present volume is, in substance, a reproduction of these lectures, with such additions and modifications as seemed to me desirable in order that the book could serve as a treatise on that part of the Calculus of Variations to which the discussion is here confined, viz., the case in which the function under the integral sign depends upon a plane curve and involves no higher derivatives than the first.

With this view I have throughout supplied the detail argumentation and introduced examples in illustration of the general principles. The emphasis lies entirely on the theoretical side: I have endeavored to give clear definitions of the fundamental concepts, sharp formulations of the problems, and rigorous demonstrations. Difficult points, such as the proof of the existence of a "field," the details in HILBERT's existence proof, etc., have received special attention.

For a rigorous treatment of the Calculus of Variations the principal theorems of the modern theory of functions of a real variable are indispensable; these I had therefore to presuppose, the more so as I deviate from WEIERSTRASS and KNESER in not assuming the function under the integral sign to be analytic. In order, however, to make the book accessible to a larger circle of readers, I have systematically given references to the following standard works: *Encyclopaedie der mathematischen Wissenschaften* (abbreviated *E.*), especially the articles on "Allgemeine Functionslehre" (PRINGSHEIM) and "Differential- und Integralrechnung" (VOSS); JORDAN, *Cours d'Analyse*, second edition (abbreviated *J.*); GENOCCHI-PEANO, *Differentialrechnung und Grundzüge der Integralrechnung*, translated by BOHLMANN and SCHEPP (abbreviated *P.*); occasionally also to DINI, *Theorie der Functionen einer veränderlichen reellen Grösse*, translated by LÜROTH and SCHEPP; STOLZ, *Grundzüge der Differential- und Integralrechnung*. The references are given for each theorem where it occurs for the first time; they may also be found by means of the index at the end of the book.

Certain developments have been given in smaller print in order to indicate, not that they are of minor importance, but that they may be passed over at a first reading and taken up only when referred to later on.

A few remarks are necessary concerning my attitude toward WEIERSTRASS's lectures. WEIERSTRASS's results and methods may at present be considered as generally known, partly through dissertations and other publications of his pupils, partly through KNESER's *Lehrbuch der Variationsrechnung* (Braunschweig, 1900), partly through sets of notes ("Ausarbeitungen") of which a great number are in circulation and copies of which are accessible to everyone in the library of the Mathematische Verein at Berlin, and in the Mathematische Lesezimmer at Göttingen.

Under these circumstances I have not hesitated to make use of WEIERSTRASS's lectures just as if they had been published in print.

My principal source of information concerning WEIERSTRASS's theory has been the course of lectures on the Calculus of Variations of the Summer Semester, 1879, which I had the good fortune to attend as a student in the University of Berlin. Besides, I have had at my disposal sets of notes of the courses of 1877 (by MR. G. SCHULZ) and of 1882 (a copy of the set of notes in the "Lesezimmer" at Göttingen for which I am indebted to PROFESSOR TANNER), a copy of a few pages of the course of 1872 (from notes taken by MR. OTT), and finally a set of notes (for which I am indebted to DR. J. C. FIELDS) of a course of lectures on the Calculus of Variations by PROFESSOR H. A. SCHWARZ (1898–99).

I regret very much that I have not been able to make use of the articles on the Calculus of Variations in the *Encyclopaedie der mathematischen Wissenschaften* by KNESER, ZERMELO, and HAHN. When these articles appeared, the printing of this volume was practically completed. For the same reason no reference could be made to HANCOCK's *Lectures on the Calculus of Variations.*

In concluding, I wish to express my thanks to PROFESSOR G. A. BLISS for valuable suggestions and criticisms, and to DR. H. E. JORDAN for his assistance in the revision of the proof-sheets.

OSKAR BOLZA.

THE UNIVERSITY OF CHICAGO,
August 28, 1904.

TABLE OF CONTENTS

CHAPTER I

CHAPTER II

CHAPTER III

CHAPTER IV

WEIERSTRASS'S THEORY OF THE PROBLEM IN PARAMETER-REPRESENTATION

CHAPTER V

KNESER'S THEORY

CHAPTER VI

WEIERSTRASS'S THEORY OF THE ISOPERIMETRIC PROBLEMS

CHAPTER VII

HILBERT'S EXISTENCE THEOREM

CHAPTER I

THE FIRST VARIATION

§1. INTRODUCTION

THE Calculus of Variations deals with problems of *maxima and minima.* But while in the ordinary theory of maxima and minima the problem is to determine those values of the independent variables for which a given function of these variables takes a maximum or minimum value, in the Calculus of Variations *definite integrals*[1] involving one or more unknown functions are considered, and it is required so to determine these unknown functions that the definite integrals shall take maximum or minimum values.

The following example will serve to illustrate the character of the problems with which we are here concerned, and its discussion will at the same time bring out certain points which are important for an exact formulation of the general problem:

EXAMPLE I: *In a plane there are given two points A, B and a straight line $\mathfrak{L}$. It is required to determine, among all curves which can be drawn in this plane between A and B, the one which, if revolved around the line $\mathfrak{L}$, generates the surface of minimum area.*

We choose the line $\mathfrak{L}$ for the x-axis of a rectangular system of co-ordinates, and denote the co-ordinates of the points A and B by x_0, y_0 and x_1, y_1 respectively. Then for a curve

$$y = f(x)$$

[1] The problem of the Calculus of Variations has, however, been extended beyond the domain of definite integrals (viz., to functions defined by differential equations) by A. MAYER, *Leipziger Berichte*, 1878 and 1895. Compare KNESER, *Lehrbuch*, chap. vii.

joining the two points A and B, the area in question is given by the definite integral[1]

$$J = 2\pi \int_{x_0}^{x_1} y\sqrt{1+y'^2}\,dx \ ,$$

where y' stands for the derivative $f'(x)$. For different curves the integral will take, in general, different values; and our problem is then analytically: among all functions $f(x)$ which take for $x = x_0$ and $x = x_1$ the prescribed values y_0 and y_1 respectively, to determine the one which furnishes the smallest value for the integral J.

This formulation of the problem implies, however, a number of tacit assumptions, which it is important to state explicitly:

a) In the first place, we must add some *restrictions concerning the nature of the functions* $f(x)$ which we admit to consideration. For, since the definite integral contains the derivative y', it is tacitly supposed that $f(x)$ has a derivative; the function $f(x)$ and its derivative must, moreover, be such that the definite integral has a determinate finite value. Indeed, the problem becomes definite only if we confine ourselves to *curves of a certain class*, characterized by a well-defined system of conditions concerning continuity, existence of derivative, etc.

For instance, we might admit to consideration only functions $f(x)$ with a continuous first derivative; or functions with continuous first and second derivatives; or analytic functions, etc.

b) Secondly, by assuming the curves *representable in the form* $y = f(x)$, where $f(x)$ is a *single-valued function* of x, we have tacitly introduced an important restriction, viz., that we consider only those curves which are met by every ordinate between x_0 and x_1 at but one point.

[1] a being a real positive quantity, $\sqrt{a}$ will always be understood to represent the positive value of the square root.

We can free ourselves from this restriction by assuming the curve in parameter-representation:[1]

$$x=\phi(t)\ , \qquad y=\psi(t)\ .$$

The integral which we have to minimize becomes then

$$J=2\pi\int_{t_0}^{t_1} y\sqrt{x'^2+y'^2}\,dt\ ,$$

where $x'=\phi'(t)$, $y'=\psi'(t)$, and where t_0 and t_1 are the values of t which correspond to the two end-points.

c) It is further to be observed that our definite integral represents the area in question only when $y\geqq 0$ throughout the interval of integration. The problem implies, therefore, the condition that *the curves shall lie in a certain region*[2] of the x, y-plane (viz., the upper half-plane).

d) Our formulation of the problem tacitly assumes that there *exists* a curve which furnishes a minimum for the area. But the existence of such a curve is by no means self-evident. We can only be sure that there exists a lower limit[3] for the values of the area; and the decision whether this lower limit is actually reached or not forms part of the solution of the problem.

The problem may be modified in various ways. For instance, instead of assuming both end-points fixed, we may assume one or both of them movable on given curves.

An essentially different class of problems is represented by the following example:

1 Compare chap. iv. Even in this generalized form the analytic problem is not quite so general as the original geometrical problem. For the area in question may exist and be finite, and yet not be representable by the above definite integral. This suggests an extension of the problem of the Calculus of Variations, first considered by WEIERSTRASS. Compare §§ 31 and 44.

2 A restriction of the same nature, but from other reasons, occurs in the problems of the brachistochrone and of the geodesic; compare § 26.

3 Compare E. I A, p. 72, and II A, p. 9; J. I, No. 25; and P., No. 20.

EXAMPLE II: *Among all closed plane curves of given perimeter to determine the one which contains the maximum area.*

If we use parameter-representation, the problem is to determine among all curves *for which the definite integral*

$$\int_{t_0}^{t_1} \sqrt{x'^2 + y'^2}\, dt$$

has a given value, the one which maximizes the integral

$$J = \tfrac{1}{2} \int_{t_0}^{t_1} (x y' - x' y)\, dt \; .$$

Here the curves out of which the maximizing curve is to be selected are subject—apart from restrictions of the kind which we have mentioned before—to the new condition of furnishing a given value for a certain definite integral. Problems of this kind are called "isoperimetric problems;" they will be treated in chap. vi.

The preceding examples are representatives of the simplest —and, at the same time, most important—type of problems of the Calculus of Variations, in which are considered definite integrals depending upon *a plane curve* and containing *no higher derivatives than the first.* To this type we shall almost exclusively confine ourselves.

The problem may be generalized in various directions:

1. Higher derivatives may occur under the integral.

2. The integral may depend upon a system of unknown functions, either independent or connected by finite or differential relations.

3. Extension to multiple integrals.

For these generalizations we refer the reader to C. JORDAN, *Cours d'Analyse*, 2e éd., Vol. III, chap. iv; PASCAL-(SCHEPP), *Die Variationsrechnung* (Leipzig, 1899); and KNESER, *Lehrbuch der Variationsrechnung* (Braunschweig, 1900), Abschnitt VI, VII, VIII.

§2. AGREEMENTS[1] CONCERNING NOTATION AND TERMINOLOGY

a) We consider exclusively *real* variables. The "*interval* $(a\ b)$" of a variable x — where the notation always implies $a<b$—is the totality of values x satisfying the inequality $a \leqq x \leqq b$. The "*vicinity* (δ) *of a point* $x_1=a_1$, $x_2=a_2, \cdots, x_n=a_n$" is the totality of points $x_1, x_2, \cdots, x_n$ satisfying the inequalities:

$$|x_1-a_1|<\delta, \quad |x_2-a_2|<\delta, \quad \cdots, \quad |x_n-a_n|<\delta .$$

The word "*domain*" will be used in the same sense as the German *Bereich*, *i. e.*, synonymous with "set of points" (compare E. II A, p. 44). The word "*region*" will be used: (*a*) for a "continuum," *i. e.*, a set of points which is "connected" and made up exclusively of "inner" points; in this case the boundary does not belong to the region ("open" region); (*b*) for a continuum together with its boundary ("closed" region); (*c*) for a continuum together with part of its boundary. The region may be finite or infinite; it may also comprise the whole n-dimensional space.

When we say: a curve lies "*in*" a region, we mean: each one of its points is a point of the region, not necessarily an inner point.

For the definition of "inner" point, "boundary point" (*frontière*), and "connected" (*d'un seul tenant*) we refer to E. II A, p. 44; J. I, Nos. 22, 31; and HURWITZ, *Verhandlungen des ersten internationalen Mathematikercongresses in Zürich*, p. 94.

b) By a "*function*" is always meant a real *single-valued* function.

The *substitution* of a particular value $x=x_0$ in a function $\phi(x)$ will be denoted by

$$\phi(x)\Big|^{x_0} = \phi(x_0) \ ;$$

[1] The reader is advised to proceed directly to §3 and to use §2 only for reference.

similarly

$$\phi(x, y)\Big|_{\substack{x=x_0\\y=y_0}} = \phi(x_0, y_0)\ ;$$

also

$$\Big[\phi(x)\Big]_{x_0}^{x_1} = \phi(x_1) - \phi(x_0)\ .$$

Instead we shall also use the simpler notation

$$\phi(x)\Big|^{0},\ \phi(x, y)\Big|^{0},\ \big[\phi(x)\big]_0^1$$

where it can be done without ambiguity, compare *e*).

We shall say: a function has a certain property IN[1] *a domain* $\mathfrak{S}$ of the independent variables, if it has the property in question at all points of the domain $\mathfrak{S}$, no matter whether they are interior or boundary points.

A function of $x_1, x_2, \cdots, x_n$ has a certain property *in the vicinity of a point* $x_1 = a_1, x_2 = a_2, \cdots, x_n = a_n$, if there exists a positive quantity δ such that the function has the property in question in the vicinity (δ) of the point $a_1, a_2, \cdots, a_n$.

If $\underset{h=0}{L}\phi(h) = 0$, we shall say: $\phi(h)$ is an "*infinitesimal*" (for $Lh = 0$); such an infinitesimal will in a general way be denoted by (h). Also an independent variable h which in the course of the investigation is made to approach zero, will be called an "infinitesimal."

c) *Derivatives* of functions of one variable will be denoted by accents, in the usual manner:

$$f'(x) = \frac{df(x)}{dx}\ ,\quad f''(x) = \frac{d^2 f(x)}{dx^2}\ ,\quad \text{etc.}$$

For brevity we shall use the following terminology[2] for various *classes of functions* which will frequently occur in the sequel. We shall say that a function $f(x)$ which is defined in an interval $(x_0 x_1)$ is

[1] Or, with more emphasis, "throughout."

[2] The letters *C*, *D* are to suggest "continuous," "*d*iscontinuous;" the accents the order of the derivative involved.

$$\left.\begin{array}{lll} \text{of class } C & \text{if } f(x) & \text{is continuous} \\ \text{of class } C' & \text{if } f(x) & \text{and } f'(x) \text{ are continuous} \\ \cdot\ \cdot\ \cdot & \cdot\ \cdot\ \cdot & \cdot\ \cdot\ \cdot \\ \text{of class } C^{(n)} & \text{if } f(x), f'(x), \cdots & \text{and } f^{(n)}(x) \text{ are continuous} \end{array}\right\} \text{ in } (x_0x_1),$$

with the understanding concerning the extremities of the interval that the definition of $f(x)$ can be so extended beyond (x_0x_1) that the above properties still hold at x_0 and x_1.

If $f(x)$ itself is continuous, and if the interval (x_0x_1) can be divided into a finite number of subintervals

$$(x_0c_1),\ (c_1c_2),\ \cdots,\ (c_{n-1}x_1),$$

such that in each subinterval $f(x)$ is of class C' (C''), whereas $f'(x)$ $(f''(x))$ is discontinuous at $c_1, c_2, \cdots, c_{n-1}$, we shall say that $f(x)$ is of class D' (D''). We consider class C' (C'') as contained in D' (D''), viz., for $n=1$.

From these definitions it follows that, for a function of class D', the progressive[1] and regressive derivatives $\overset{+}{f'}(c_\nu)$, $\overset{-}{f'}(c_\nu)$ exist, are finite and equal to the limiting values[2] $f'(c_\nu+0)$, $f'(c_\nu-0)$ respectively.

d) Partial derivatives of functions of several variables will be denoted by literal subscripts (KNESER):

$$F_y(x, y, p) = \frac{\partial F(x, y, p)}{\partial y},$$

$$F_{yp}(x, y, p) = \frac{\partial}{\partial p}\left(\frac{\partial F(x, y, p)}{\partial y}\right), \quad \text{etc.};$$

also

$$F_y(x_0, y_0, p_0) = \frac{\partial F(x, y, p)}{\partial y}\bigg|_{\substack{x=x_0\\ y=y_0\\ p=p_0}}.$$

Also of a function of several variables we shall say that it is *of class* $C^{(n)}$ in a domain $\mathfrak{S}$ if all its partial derivatives

[1] E. II A, p. 61; DINI, *Grundlagen*, etc., §68: and STOLZ, *Grundzüge*, etc., Vol. I, p. 31.

[2] E. II A, p. 13.

up to the nth order inclusive exist and are continuous in[1] the domain $\mathfrak{S}$.

e) The letters x, y will always be used for rectangular *co-ordinates* with the usual orientation of the positive axes, *i. e.*, the positive y-axis to the left of the positive x-axis. It will frequently be convenient to designate points by numbers: 0, 1, 2, $\cdots$; the co-ordinates of these points will then always be denoted by x_0, y_0; x_1, y_1; x_2, y_2; $\cdots$ respectively; their parameters, if they lie on a curve given in parameter-representation, by $t_0, t_1, t_2, \cdots$.

A *curve*[2] (arc of curve)

$$y = f(x)\ , \qquad x_0 \leqq x \leqq x_1\ ,$$

will be said to be *of class* C, C', etc., if the function $f(x)$ is of class C, C', etc., in (x_0x_1). In particular, a curve of class D' is continuous and made up of a finite number of arcs with continuously turning tangents, not parallel to the y-axis. The points of the curve whose abscissæ are the points of discontinuity C_1, C_2, $\cdots$, C_{n-1} of $f'(x)$, $\cdots$ will be called its *corners*. At a corner the curve has a progressive and a regressive tangent, and,

$$\tan\overset{+}{\alpha} = \overset{+}{f}{}'(c)\ , \quad \tan\overset{-}{\alpha} = \overset{-}{f}{}'(c)\ .$$

(See Fig. 1.)

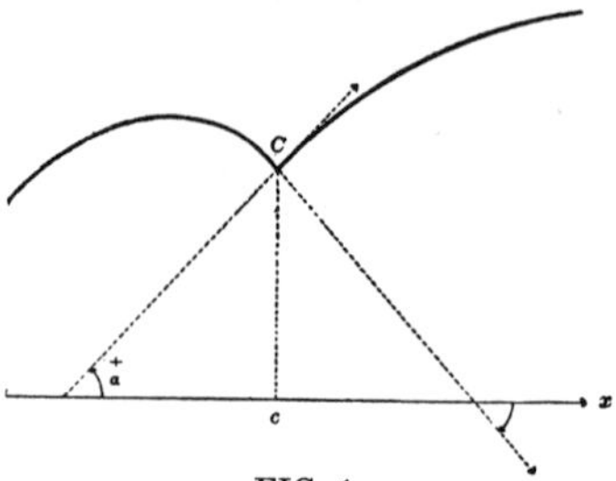

FIG. 1

f) The *integral*

$$J = \int F(x, y, y')\, dx$$

taken along the curve

$$\mathfrak{C}: \qquad y = f(x)\ , \qquad x_0 \leqq x \leqq x_1$$

[1] When $\mathfrak{S}$ contains boundary points, an agreement similar to that in the case of one variable is necessary with respect to these points.

[2] The corresponding definitions for curves in parameter-representation will be given in §24.

from the point $A(x_0, y_0)$ to the point $B(x_1, y_1)$, *i. e.*, the integral

$$\int_{x_0}^{x_1} F\left(x, f(x), f'(x)\right) dx$$

will be denoted by $J_{\mathfrak{C}}(AB)$ (more briefly $J_{\mathfrak{C}}$ or $J(AB)$); or by $J_{\mu\nu}$, if the end-points are designated by numbers: μ, ν.

g) The *distance* between the two points P and Q will be denoted by $|PQ|$, the *circle* with center O and radius r by (O, r) (HARKNESS AND MORLEY). The angle which a vector makes with the positive x-axis will be called its *amplitude.*

§3. GENERAL FORMULATION OF THE PROBLEM[1]

a) After these preliminary explanations, the simplest problem of the Calculus of Variations may be formulated in the most general way, as follows:

There is given:

1. A well-defined infinitude $\mathfrak{M}$ of curves, representable in the form

$$y = f(x) \ , \qquad x_0 \leqq x \leqq x_1 \ ;$$

the end-points and their abscissæ x_0, x_1 may vary from curve to curve. We shall refer to these curves as "admissible curves."

2. A function $F(x, y, p)$ of three independent variables such that for every admissible curve $\mathfrak{C}$, the definite integral

$$J_{\mathfrak{C}} = \int_{x_0}^{x_1} F(x, y, y')\, dx \tag{1}$$

has a determinate finite value.

[1] Until rather recently a certain vagueness has prevailed with respect to the fundamental concepts of the Calculus of Variations. The most important contributions toward clear definitions and sharp formulations of the problems are due to DU BOIS-REYMOND, "Erläuterungen zu den Anfangsgründen der Variationsrechnung," *Mathematische Annalen*, Vol. XV (1879), p. 283; SCHEEFFER, "Ueber die Bedeutung der Begriffe 'Maximum und Minimum' in der Variationsrechnung," *ibid.*, Vol. XXVI (1886), p. 197; WEIERSTRASS, *Lectures on the Calculus of Variations*, especially those since 1879. Compare also ZERMELO, *Untersuchungen zur Variationsrechnung, Dissertation* (Berlin, 1894), p. 24; KNESER, *Lehrbuch*, §17, and OSGOOD, "Sufficient Conditions in the Calculus of Variations," *Annals of Mathematics* (2), Vol. II (1901), p. 105.

The set[1] of values $J_{\mathfrak{C}}$ thus defined has always a lower limit, K, and an upper limit, G (finite or infinite[2]). If the lower (upper) limit is finite, and if there exists an admissible curve $\mathfrak{C}$ such that

$$J_{\mathfrak{C}} = K \ , \qquad (J_{\mathfrak{C}} = G) \ ,$$

the curve $\mathfrak{C}$ is said to furnish *the absolute minimum* (*maximum*) for the integral J (with respect to $\mathfrak{M}$). For every other admissible curve $\overline{\mathfrak{C}}$ we have then

$$J_{\overline{\mathfrak{C}}} \geqq J_{\mathfrak{C}} \ , \qquad (J_{\overline{\mathfrak{C}}} \leqq J_{\mathfrak{C}}) \ . \tag{2}$$

The word "extremum"[3] will be used for maximum and minimum alike, when it is not necessary to distinguish between them.

Hence the *problem* arises: to determine all admissible curves which, in this sense, minimize or maximize the integral J.

b) As in the theory of ordinary maxima and minima, the problem of the absolute extremum, which is the ultimate aim of the Calculus of Variations, is reducible to another problem which can be more easily attacked, viz., the problem of the relative extremum:

An admissible curve $\mathfrak{C}$ is said to furnish a *relative minimum*[4] (*maximum*) if there exists a "*neighborhood* $\mathfrak{U}$ *of the curve* $\mathfrak{C}$," however small, such that the curve $\mathfrak{C}$ furnishes an absolute minimum with respect to the totality $\mathfrak{M}_1$ of those curves of $\mathfrak{M}$ which lie in this neighborhood; and by a neighborhood $\mathfrak{U}$ of the curve $\mathfrak{C}$ we understand any region[5] which contains $\mathfrak{C}$ in its *interior*.

[1] By "set" we translate the German *Punktmenge*, the French *ensemble*, J. I, No. 20.

[2] The upper limit is $+\infty$, if for every preassigned positive quantity A there exist curves $\mathfrak{C}$ for which $J_{\mathfrak{C}} > A$; see E. II A, p. 9.

[3] DU BOIS-REYMOND, *Mathematische Annalen*, Vol. XV, p. 564.

[4] In the use of the words "absolute" and "relative" I follow Voss in E. II A, p. 80. Many authors call the isoperimetric problems "problems of relative maxima and minima."

[5] For the definition of the term "region," see p. 5.

According to STOLZ, the relative minimum (maximum) will be called *proper*, if there exists a neighborhood $\mathfrak{U}$ such that in (2) the sign $>$ ($<$) holds for all curves $\overline{\mathfrak{C}}$ different from $\mathfrak{C}$; *improper* if, however the neighborhood $\mathfrak{U}$ may be chosen, there exists some curve $\overline{\mathfrak{C}}$ different from $\mathfrak{C}$ for which the equality sign has to be taken.

A curve which furnishes an absolute extremum evidently furnishes *a fortiori* also a relative extremum. Hence the original problem is reducible[1] to the problem: *to determine all those curves which furnish a relative minimum;* and in this form we shall consider the problem in the sequel.

We shall henceforth always use the words "minimum," "maximum" in the sense of relative minimum, maximum; and we shall confine ourselves to the case of a minimum, since every curve which minimizes J, at the same time maximizes $-J$, and *vice versa.*

c) In the abstract formulation given above, the problem would hardly be accessible to the methods of analysis; to make it so, it is necessary to specify some concrete assumptions concerning the admissible curves and the function F.

For the present, we shall make the following assumptions:

A. The infinitude $\mathfrak{M}$ of admissible curves shall be the totality of all curves satisfying the following conditions:

1. They pass through two given points $A(x_0, y_0)$ and $B(x_1, y_1)$.

2. They are representable in the form

$$y = f(x) \;, \qquad x_0 \leqq x \leqq x_1 \;,$$

$f(x)$ being a single-valued function of x.

3. They are *continuous and consist of a finite number of*

[1] After the relative problem has been solved, it merely remains to pick out among its solutions those which furnish the smallest or largest value for J. Only if the relative problem should have an infinitude of solutions, new difficulties would arise. For a direct treatment of the problem of the absolute extremum compare HILBERT'S existence proof (chap. vii); DARBOUX, *Théorie des surfaces*, Vol. III, p. 89; and ZERMELO, *Jahresbericht der Deutschen Mathematiker-Vereinigung*, Vol. XI (1902), p. 184.

arcs with continuously turning tangents, not parallel to the y-axis; i. e., in the terminology of §2, c), $f(x)$ is of class D'.

4. They lie in a given region[1] $\mathfrak{R}$ of the x, y-plane.

B. The function $F(x, y, p)$ shall be continuous[2] and admit continuous partial derivatives of the first, second, and third orders in a domain[3] $\mathfrak{T}$ which consists of all points[4] (x, y, p) for which (x, y) is a point of $\mathfrak{R}$, and p has a finite value.

Under these assumptions the definite integral $J_{\mathfrak{C}}$ taken along any admissible curve $\mathfrak{C}$ is always finite and determinate,[5] provided we define, in the case of a curve with corners, the integral as the sum of integrals taken between two successive corners. Since we suppose the end-points A and B fixed and the curves representable in the form $y=f(x)$, the curves $\mathfrak{C}$ all lie between the two lines $x=x_0$ and $x=x_1$, with the exception of the end-points, which lie on these lines.

Hence it follows that we may, in the present case, give the following simpler definition of a minimum: An admissible curve $\mathfrak{C}: y=f(x)$ minimizes the integral J, if[6] there

[1] Compare §2, a).

[2] I follow here the example of PASCAL, *loc. cit.*, p. 21, and OSGOOD, *loc. cit.*, p. 105. WEIERSTRASS, JORDAN, and KNESER suppose the function $F(x, y, p)$ to be *analytic*.

[3] If we interpret p as a third co-ordinate perpendicular to the x, y-plane, $\mathfrak{T}$ is the cylinder, infinite in both directions, whose base is the region $\mathfrak{R}$.

[4] "Point" in the sense of the theory of "point-sets." Compare E. II A, p. 44, and J. I, No. 20.

[5] If the curve has no corners, this follows at once from elementary theorems on continuous functions (J. I, Nos. 60, 66). If the curve has corners, the integral $J_{\mathfrak{C}}$ has no immediate meaning. But the two integrals

$$\int_{x_0}^{x_1} F\left(x, f(x), \overset{+}{f'}(x)\right) dx \quad \text{and} \quad \int_{x_0}^{x_1} F\left(x, f(x), \overset{-}{f'}(x)\right) dx$$

are finite and determinate and equal to each other, and at the same time equal to the sum of integrals mentioned in the text. Compare DINI, *loc. cit.*, §62; §187, 2; §190, 9; and §190, 2.

[6] In admitting the equality sign in the inequality (2), I deviate from the conventions generally adopted in the Calculus of Variations and follow STOLZ (*Grundzüge der Differenzialrechnung*, Vol. I, p. 199), whose definition is more consistent with the usual definition of absolute minimum. If the equality sign were omitted, it could not be said that every curve which furnishes an absolute minimum furnishes *a fortiori* also a relative minimum.

exists a positive quantity ρ such that $J_{\bar{\mathfrak{C}}} \geqq J_{\mathfrak{C}}$ for every admissible curve $\bar{\mathfrak{C}}: \bar{y} = \bar{f}(x)$ which satisfies the inequality

$$|\bar{y} - y| < \rho \quad \text{for} \quad x_0 \leqq x \leqq x_1 \ . \tag{3}$$

This means geometrically that the curve $\bar{\mathfrak{C}}$ lies in the interior[1] of the strip of the x, y-plane between the two curves

$$y = f(x) + \rho \ , \qquad y = f(x) - \rho$$

on the one hand, and the two lines $x = x_0$, $x = x_1$ on the other hand. This strip we shall call "the neighborhood[2] (ρ) of the curve $\mathfrak{C}$," the points A and B being included, the rest of the boundary excluded.

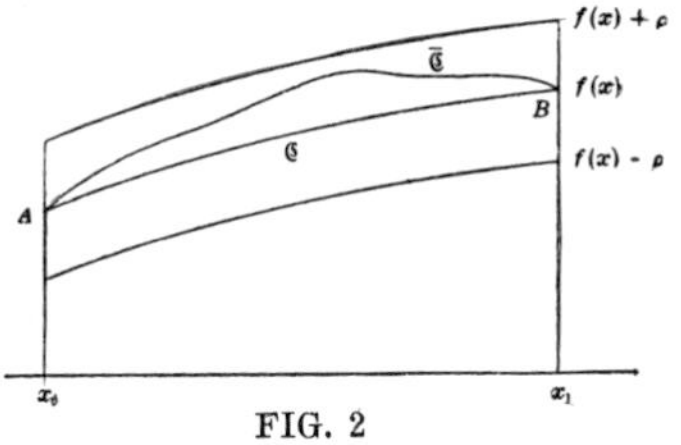

FIG. 2

§4. VANISHING OF THE FIRST VARIATION

We now suppose we have found a curve $\mathfrak{C}: y = f(x)$ which minimizes the integral

$$J = \int_{x_0}^{x_1} F(x, y, y')\, dx$$

in the sense explained in the last section. We further suppose, for the present,[3] that $f'(x)$ *is continuous* in $(x_0 x_1)$ and that $\mathfrak{C}$ lies entirely in the *interior* of the region $\mathfrak{R}$.

From the last assumption it follows that we can construct[4] a neighborhood (ρ) of $\mathfrak{C}$ which lies entirely in the interior of $\mathfrak{R}$.

1 Except, of course, the points A and B.

2 Compare OSGOOD, *loc. cit.*, p. 107.

3 These restrictions will be dropped in §§9 and 10.

4 About any point P of $\mathfrak{C}$ we can construct a circle (P, r) which lies entirely in $\mathfrak{R}$, since P is an inner point of $\mathfrak{R}$. Let ρ_P be the upper limit of the values of r for which this takes place. Then ρ_P varies continuously as P describes the curve $\mathfrak{C}$ (WEIERSTRASS, *Werke*, Vol. II, p. 204) and reaches therefore a positive minimum value ρ_0 (compare E. II A, p. 19 and J. I, No. 64, Cor.). If we choose $\rho < \rho_0$ the neighborhood (ρ) of $\mathfrak{C}$ will lie in the interior of $\mathfrak{R}$.

We then replace[1] the curve $\mathfrak{C}$ by another admissible curve

$$\overline{\mathfrak{C}}: \qquad \bar{y} = \bar{f}(x) \ ,$$

lying entirely in the neighborhood (ρ).

The increment

$$\Delta y = \bar{y} - y = \bar{f}(x) - f(x) \ ,$$

which we shall denote by ω, is called the *total variation of y*. Since $\mathfrak{C}$ and $\overline{\mathfrak{C}}$ pass through A and B, we have

$$\omega(x_0) = 0 \ , \qquad \omega(x_1) = 0 \ , \tag{4}$$

and since $\overline{\mathfrak{C}}$ lies in (ρ),

$$|\omega(x)| < \rho \qquad \text{in } (x_0 x_1) \ . \tag{4a}$$

The corresponding increment of the integral,

$$\Delta J = J_{\overline{\mathfrak{C}}} - J_{\mathfrak{C}} \ ,$$

is called the *total variation of the integral J*; it may be written:

$$\Delta J = \int_{x_0}^{x_1} \Big[F(x, y + \omega, y' + \omega') - F(x, y, y') \Big] dx \ .$$

Since $\mathfrak{C}$ is supposed to minimize J, we shall have

$$\Delta J \geqq 0 \ ,$$

provided that ρ has been chosen sufficiently small.

For the next step in the discussion of this inequality two different methods have been proposed:

a) *Application of Taylor's formula*: If we apply Taylor's[2] formula to the integrand of ΔJ, we obtain, in the notation of §2, *d*),

[1] The process of replacing $\mathfrak{C}$ by $\overline{\mathfrak{C}}$ is called "*a variation of the curve* $\mathfrak{C}$;" the same term is frequently applied to the curve $\overline{\mathfrak{C}}$ itself, which is sometimes also called "the varied curve," or "a neighboring curve."

[2] The conditions for the applicability of Taylor's formula are fulfilled, compare E. II A, p. 77, and J. I, No. 253. $F_{y'}$, $F_{yy'}$, etc., are synonymous with F_p, F_{yp}, etc. The method here used was first given by LAGRANGE. See *Oeuvres*, Vol. IX, p. 297. Compare also DU BOIS-REYMOND, *Mathematische Annalen*, Vol. XV (1879), p. 292, and PASCAL-SCHEPP, *Die Variationsrechnung*, p. 22.

Instead of Taylor's formula with the remainder-term, WEIERSTRASS (*Lectures*), KNESER (*Lehrbuch der Variationsrechnung*, §8), and C. JORDAN (*Cours d'Analyse*, Vol. III, No. 350), who suppose $F(x, y, p)$ to be analytic, use Taylor's expansion into an infinite series. Here, however, the question of integration by terms should be considered.

$$\Delta J = \int_{x_0}^{x_1} (F_y \omega + F_{y'} \omega') \, dx$$
$$+ \tfrac{1}{2} \int_{x_0}^{x_1} (\widetilde{F}_{yy} \omega^2 + 2 \widetilde{F}_{yy'} \omega\omega' + \widetilde{F}_{y'y'} \omega'^2) \, dx \; ,$$

where the arguments of F_y and $F_{y'}$ are x, y, y', those of $\widetilde{F}_{yy}$, $\widetilde{F}_{yy'}$, $\widetilde{F}_{y'y'}$: x, $y + \theta\omega$, $y' + \theta\omega'$, θ being a quantity between 0 and 1.

We now consider, with LAGRANGE,[1] special[2] variations of the form

$$\omega = \epsilon \eta \; , \tag{5}$$

where η is a function of x of class D' which vanishes for $x = x_0$ and $x = x_1$, and ϵ a constant whose absolute value is taken so small that (4a) is satisfied.

Then ΔJ takes the form[3]

$$\Delta J = \epsilon \left\{ \int_{x_0}^{x_1} (F_y \eta + F_{y'} \eta') \, dx + (\epsilon) \right\} \; , \tag{6}$$

where (ϵ) denotes an infinitesimal for $L\epsilon = 0$.

Hence we infer that we must have

$$\int_{x_0}^{x_1} (F_y \eta + F_{y'} \eta') \, dx = 0 \tag{7}$$

for all functions η of class D$'$ which vanish at x_0 and x_1;

[1] *Oeuvres*, Vol. IX, p. 298.

[2] For the purpose of deriving necessary conditions, we may specialize the variations as much as convenient. It will be different when we come to sufficient conditions (compare §17).

[3] *Proof:* We suppose first that $\eta'(x)$ is continuous in $(x_0 x_1)$ and denote by μ and μ' the maxima of $|\eta(x)|$ and $|\eta'(x)|$ in $(x_0 x_1)$, and by q a quantity greater than the maximum of $|f'(x)|$ in $(x_0 x_1)$. Having once chosen the function $\eta(x)$, we can then determine a positive quantity δ such that the point $(x, \overline{y})$ lies in the neighborhood (ρ) of $\mathfrak{C}$ and that $-q < \widetilde{y}' < q$ for every x in $(x_0 x_1)$, provided that $|\epsilon| < \delta$. On the other hand, the three functions $|\widetilde{F}_{yy}|$, $|\widetilde{F}_{yy'}|$, $|\widetilde{F}_{y'y'}|$ remain, in this domain, below a finite fixed quantity G. Hence, by the mean-value theorem,

$$\left| \int_{x_0}^{x_1} (\widetilde{F}_{yy} \omega^2 + 2 \widetilde{F}_{yy'} \omega\omega' + \widetilde{F}_{y'y'} \omega'^2) \, dx \right| \leqq \epsilon^2 G (\mu^2 + 2\mu\mu' + \mu^2)(x_1 - x_0) \; .$$

If $\eta'(x)$ is not continuous in $(x_0 x_1)$, apply the same reasoning to the integrals taken between two successive corners of $\overline{\mathfrak{C}}$.

for otherwise we could make ΔJ negative as well as positive by giving ϵ once negative and once positive sufficiently small values.

b) Differentiation with respect to ϵ: The same result (7) as well as formula (6) can be obtained by the remark, due to LAGRANGE,[1] that by the substitution of $\epsilon\eta$ for ω, the integral $\bar{J}$ becomes a function of ϵ, say $J(\epsilon)$, which must have a minimum for $\epsilon=0$. Hence we must have[2] $J'(0)=0$. If $\eta(x)$ is of class C' in $(x_0 x_1)$, it follows from our assumptions concerning the function F and the curve $\mathfrak{C}$ that

$$\frac{\partial F\big(x,\, y(x)+\epsilon\eta(x),\, y'(x)+\epsilon\eta'(x)\big)}{\partial\epsilon}$$

is a continuous function of x and ϵ in the domain,

$x_0 \leqq x \leqq x_1$, $|\epsilon| \leqq \epsilon_0$, ϵ_0 being a sufficiently small positive quantity, and therefore the ordinary rule[3] for the differentiation of a definite integral with respect to a parameter may be applied. Hence we obtain

$$\left.\frac{dJ(\epsilon)}{d\epsilon}\right|^{\epsilon=0} = \int_{x_0}^{x_1} (F_y\eta + F_{y'}\eta')\,dx\ .$$

This proves (7) and at the same time (6), since by the definition of the derivative,

$$\Delta J = J(\epsilon) - J(0) = \epsilon\big(J'(0) + (\epsilon)\big)\ .$$

If $\eta(x)$ is of class D', decompose the integral $\bar{J}$ in the manner described in §3, *c*), and then proceed as above.

c) The symbol δ: We now make use of the following permanent notation introduced by LAGRANGE[4] (1760).

Let $\phi(x, y, y', y'', \cdots)$ be a function of x, y and some of the derivatives of y, whose partial derivatives with respect

[1] *Oeuvres*, Vol. X, p. 400. This method has been adopted by LINDELÖF-MOIGNO, DIENGER, and OSGOOD.

[2] Moreover $J''(0)$ must be $\geqq 0$. This condition will be discussed in chap. ii.

[3] Compare E. II A, p. 102; J. I, No. 83.

[4] *Oeuvres*, Vol. I, p. 336. Compare also J. III, No. 348.

to $y, y', y'', \cdots$ up to the nth order exist and are continuous in a certain domain. Then if we replace y by $\bar{y}=y+\epsilon\eta$, and accordingly y' by $\bar{y}'=y'+\epsilon\eta'$, etc., we can expand the function

$$\bar{\phi}=\phi(x, y+\epsilon\eta, y'+\epsilon\eta', \cdots)$$

according to powers of ϵ and obtain an expansion of the form

$$\bar{\phi}=\phi+\frac{\epsilon}{1}\phi_1+\frac{\epsilon^2}{2!}\phi_2+\cdots+\frac{\epsilon^n}{n!}\phi_n+\epsilon^n(\epsilon),$$

where (ϵ) denotes as usual an infinitesimal, and

$$\phi_1=\phi_y\eta+\phi_{y'}\eta'+\phi_{y''}\eta''+\cdots$$
$$\phi_2=\phi_{yy}\eta^2+\phi_{y'y'}\eta'^2+\cdots+2\phi_{yy'}\eta\eta'+\cdots$$
$$\cdots\cdots\cdots\cdots\cdots\cdots\cdots$$

The quantities $\epsilon\phi_1, \epsilon^2\phi_2, \cdots$ are called *the first, second, $\cdots$ variation of* ϕ and are denoted by $\delta\phi, \delta^2\phi, \cdots$ respectively.

It is easily seen that

$$\delta^k\phi=\delta(\delta^{k-1}\phi).$$

Again, if ϕ does not contain ϵ, $\delta^k\phi$ may be defined by

$$\delta^k\phi=\left.\frac{\partial^k\bar{\phi}}{\partial\epsilon^k}\right|^{\epsilon=0}\cdot\epsilon^k.$$

Similarly, $\delta^k J$ is defined as the term of order k, multiplied by $k!$, in the expansion of

$$\bar{J}=\int_{x_0}^{x_1}F(x, y+\epsilon\eta, y'+\epsilon\eta')\,dx$$

according to powers of ϵ, the possibility of this expansion up to terms of order k being, of course, presupposed. Accordingly

$$\delta^k J=\left.\frac{d^k\bar{J}}{d\epsilon^k}\right|^{\epsilon=0}\cdot\epsilon^k.$$

It follows immediately[1] that

$$\delta^k J = \int_{x_0}^{x_1} \delta^k F \, dx \; .$$

In particular

$$\delta J = \epsilon \int_{x_0}^{x_1} (F_y \eta + F_{y'} \eta') \, dx \; . \tag{8}$$

We may therefore formulate the result reached above as follows: *For an extremum it is necessary that the first variation of the integral J shall vanish for all admissible variations of the function y.*

d) More general type of variations: For many investigations it is necessary to extend the important formula (6) to variations of the following more general type:[2]

$$\omega = \omega(x, \epsilon) \; , \tag{5a}$$

where $\omega(x, \epsilon)$ is a function of x and ϵ which vanishes identically for $\epsilon = 0$. We suppose that $\omega(x, \epsilon)$ together with the partial derivatives ω_x, ω_ϵ, $\omega_{x\epsilon}$ are continuous in the domain

$$x_0 \leqq x \leqq x_1 \; , \qquad |\epsilon| \leqq \epsilon_0 \; ,$$

ϵ_0 being a sufficiently small positive quantity.

Moreover, in the case when both end-points are fixed

$$\omega(x_0, \epsilon) = 0 \qquad \text{and} \qquad \omega(x_1, \epsilon) = 0$$

for every $|\epsilon| \leqq \epsilon_0$. If we denote $\omega_\epsilon(x, 0)$ by $\eta(x)$, formula (6) holds also for variations of type (5a). This can be most easily proved by the method explained under b).

For the function

$$J(\epsilon) = \int_{x_0}^{x_1} F\Big(x, \, y(x) + \omega(x, \epsilon), \, y'(x) + \omega_x(x, \epsilon)\Big) \, dx$$

must have a minimum for $\epsilon = 0$, and therefore $J'(0) = 0$. From the above assumptions concerning $\omega(x, \epsilon)$ it follows that differentiation under the sign is allowed and that $\omega_{\epsilon x}$ exists and is equal[3] to $\omega_{x\epsilon}$.

[1] Provided always that the limits are fixed and that the ordinary rules for the differentiation of a definite integral with respect to a parameter are applicable.

[2] Such variations were already considered by LAGRANGE, *Oeuvres*, Vol. X, p. 400.

[3] Compare E. II A, p. 73.

Hence we obtain[1] also in the present case

$$J'(0) = \int_{x_0}^{x_1} (F_y \eta + F_{y'} \eta')\, dx \ ,$$

which leads immediately to (6).

For variations of type (5a) the definition of the symbol δ must be modified. In order to cover also the case of variable end-points, we suppose that $\bar{x}_0$ and $\bar{x}_1$ are functions of ϵ which reduce to x_0 and x_1 respectively, for $\epsilon = 0$. Putting then as before

$$\bar{y} = y(x) + \omega(x, \epsilon) \ , \qquad \bar{y}' = y'(x) + \omega_x(x, \epsilon) \ ,$$

we define[2]

$$\delta^k J = \frac{d^k}{d\epsilon^k} \int_{\bar{x}_0}^{\bar{x}_1} F(x, \bar{y}, \bar{y}')\, dx \Bigg|^{\epsilon = 0} \cdot \epsilon^k \ ,$$

and similarly if ϕ is a function of $x, y, y', \cdots$ and $\bar{x}_0, \bar{x}_1$,

$$\delta^k \phi = \frac{\partial^k \phi(x, \bar{y}, \bar{y}', \cdots, \bar{x}_0, \bar{x}_1)}{\partial \epsilon^k} \Bigg|^{\epsilon = 0} \cdot \epsilon^k \ .$$

The definition of the symbol δ given under *b*) is a special case of this general definition.

The method of differentiation with respect to ϵ, especially when combined with the consideration of variations of type (5a), seems to reduce the problem of the Calculus of Variations to a problem of the theory of ordinary maxima and minima; only apparently, however; for, as will be seen later, the method furnishes only necessary

[1] For variations of the special type (5) equation (6) may also be written

$$\Delta J = \int_{x_0}^{x_1} (F_y \omega + F_{y'} \omega')\, dx + \epsilon(\epsilon) \ . \tag{6a}$$

This formula remains true for variations of the more general type (5a). For from the properties of $\omega(x, \epsilon)$ it follows that the quotients

$$\big(\omega(x, \epsilon) - \omega(x, 0)\big)/\epsilon \qquad \text{and} \qquad \big(\omega_x(x, \epsilon) - \omega_x(x, 0)\big)/\epsilon$$

approach for $L\, \epsilon = 0$ their respective limits $\omega_\epsilon(x, 0)$ and $\omega_{x\epsilon}(x, 0)$ *uniformly* for all values of x in the interval $(x_0 x_1)$ (compare E. II A, pp. 18, 49, 52, 65; J. I, Nos. 62, 78 and P., Nos. 45, 100). Hence it follows that

$$\int_{x_0}^{x_1} (F_y \omega + F_{y'} \omega')\, dx = \epsilon \int_{x_0}^{x_1} (F_y \eta + F_{y'} \eta')\, dx + \epsilon(\epsilon) \ ,$$

which proves the above statement.

[2] Always under the assumption that all the derivatives occurring in the process exist and are continuous.

conditions, but is inadequate for the discussion of sufficient conditions, whereas the method based upon Taylor's formula, though less elegant, furnishes not only necessary but also sufficient conditions, at least for a so-called weak minimum (compare §17, *b*).

e) Transformation of the first variation by integration by parts:

For the further discussion of equation (7) it is customary to integrate the second term of δJ by parts:

$$\delta J = \epsilon \left\{ \Big[\eta F_{y'} \Big]_{x_0}^{x_1} + \int_{x_0}^{x_1} \eta \left(F_y - \frac{d}{dx} F_{y'} \right) dx \right\} . \tag{9}$$

Since η vanishes at x_0 and x_1, this leads to the result that for an extremum it is necessary that

$$\int_{x_0}^{x_1} \eta \left(F_y - \frac{d}{dx} F_{y'} \right) dx = 0 \tag{10}$$

for all functions η of class D' which vanish at x_0 and x_1.

The integration by parts presupposes, however, that not only y' but also *y'' exists and is continuous in (x_0x_1)*, and for the present we shall make this further restricting assumption[1] concerning the minimizing curve.

§5. THE FUNDAMENTAL LEMMA AND EULER'S EQUATION

To derive further conclusions from the last equation we need the following theorem, which is known as the Fundamental Lemma of the Calculus of Variations:

If M is a function of x which is continuous in (x_0x_1), and if

$$\int_{x_0}^{x_1} \eta M \, dx = 0 \tag{11}$$

[1] The necessity of this assumption was first emphasized by DU BOIS-REYMOND in the paper referred to on p. 9). If y'' does not exist, the existence of $\frac{d}{dx} F_{y'}$ becomes doubtful. The restriction will be dropped in §6. Discontinuities of η' of the kind here admitted do not interfere with the above results (9) and (10), since η itself is continuous. For the principles involved in the integration by parts, compare E. II A, p. 99, and J. I, Nos. 81, 84.

for all functions η which vanish at x_0 and x_1 and which admit a continuous derivative in (x_0x_1), then

$$M = 0 \tag{12}$$

in (x_0x_1).

For suppose $M(x') \neq 0$, say > 0, at a point x' of the interval (x_0x_1); then we can, on account[1] of the continuity of M, assign a subinterval $(\xi_0\xi_1)$ of (x_0x_1) containing x' and such that $M > 0$ throughout $(\xi_0\xi_1)$. Now choose $\eta \equiv 0$ outside of $(\xi_0\xi_1)$ and $\eta = (x-\xi_0)^2(x-\xi_1)^2$ in $(\xi_0\xi_1)$; this function admits a continuous derivative in (x_0x_1), vanishes at x_0 and x_1, and nevertheless makes

$$\int_{x_0}^{x_1} \eta M \, dx > 0 ,$$

contrary to the hypothesis (11); therefore $M(x') \neq 0$ is impossible.[2]

The conditions of this lemma are fulfilled for equation (10); for, since we suppose y'' to exist and to be continuous in (x_0x_1), the function

$$M = F_y - \frac{d}{dx} F_{y'}$$

is continuous[3] in (x_0x_1).

[1] Compare P., No. 17.

[2] This proof is due to DU BOIS-REYMOND (*Mathematische Annalen*, Vol. XV (1879), pp. 297, 300). In the same paper he proves that the conclusion $M = 0$ remains valid even if the equation (11) is known to hold only:

1. For all functions η having continuous derivatives up to the nth order, inclusive: proceed as above and choose, for $(\xi_0\xi_1)$,

$$\eta = (x-\xi_0)^{n+1} (\xi_1 - x)^{n+1} ;$$

2. For all functions having *all* their derivatives continuous.

H. A. SCHWARZ goes still farther and proves the conclusion valid if the η's are supposed *regular* in (x_0x_1), *i. e.*, developable into ordinary power series $\mathfrak{P}(x-x')$ in the vicinity of every point x' of the interval (x_0x_1) *Lectures on the Calculus of Variations*, Berlin, 1898–99, unpublished.)

On the other hand, the proof given in most text-books, in which

$$\eta = (x-x_0)(x_1-x) M$$

is used, assumes that (11) holds for all continuous functions η vanishing at x_0, x_1, or else, if the assumptions of the lemma concerning η are not changed, that M' exists and is continuous. This last assumption would, in our case, imply that y''' exists and is continuous.

Also HEINE'S proof (*Mathematische Annalen*, Vol. II (1870), p. 189) could be applied to our case only after further restricting assumptions concerning y.

[3] Compare J. I, No. 60, and P., No. 99.

Hence we obtain the *first necessary condition* for an extremum:

FUNDAMENTAL THEOREM I:[1] *Every function y which minimizes or maximizes the integral*

$$J = \int_{x_0}^{x_1} F(x, y, y')\, dx$$

must satisfy the differential equation

$$F_y - \frac{d}{dx} F_{y'} = 0 \ . \tag{I}$$

This differential equation was first discovered by EULER[2] in 1744, and will be referred to as *Euler's (differential) equation.*[3]

§6. DU BOIS-REYMOND'S AND HILBERT'S PROOFS OF EULER'S EQUATION

The preceding method, which was based upon the integration by parts of §4, furnishes only those solutions of our problem which admit a continuous second derivative. The question arises: Do there exist any other solutions and if so, how can we find them?

In order to answer this question, we return to the equation $\delta J = 0$ in the original form (7) and, with DU BOIS-REYMOND and HILBERT, *integrate the first*, instead of the second, *term by parts.* Since η vanishes at both end-points, we get:

$$\int_{x_0}^{x_1} \eta' \left(F_{y'} - \int_{x_0}^{x} F_y\, dx\right) dx = 0 \ . \tag{13}$$

1 We have *proved* this theorem only for functions y having a continuous second derivative. The extension to functions having only a continuous first derivative follows in §6, to functions of class D' in §9.

2 EULER, *Methodus inveniendi lineas curvas maximi minimive proprietate gaudentes*, chap. ii, art. 21; in STÄCKEL's translation in OSTWALD's *Klassiker der exakten Wissenschaften*, No. 46, p. 54.

3 HILBERT, and others call it "Lagrange's Equation." LAGRANGE himself attributes it to EULER. See *Oeuvres de Lagrange*, Vol. X, p. 397: "cette équation est celle qu'EULER a trouvée le premier."

This integration by parts is legitimate, even if y'' should not exist, since it presupposes only the continuity[1] of F_y and η'.

We are thus led to the problem:

If $N(x)$ be continuous in (x_0x_1), and if

$$\int_{x_0}^{x_1} \eta' N\, dx = 0 \tag{14}$$

for all functions η of class C' which vanish at x_0 and x_1, what follows with respect to N?

The answer is that N *must be constant in* (x_0x_1).

a) DU BOIS-REYMOND[2] reaches this result by the following device:

Let ζ be any function which is continuous in (x_0x_1) and satisfies the condition

$$\int_{x_0}^{x_1} \zeta\, dx = 0 \; ; \tag{15}$$

then the function

$$\eta = \int_{x_0}^{x} \zeta\, dx$$

is of class C' in (x_0x_1) and vanishes for $x = x_0$ and $x = x_1$, and therefore, according to our hypothesis, satisfies (14), that is,

$$\int_{x_0}^{x_1} \zeta N\, dx = 0 \; . \tag{16}$$

Thus it follows from our hypothesis that every continuous function which satisfies (15) necessarily satisfies (16) also.

Now let ζ_1 be any continuous function of x; and c the following constant:

$$c = \frac{1}{x_1 - x_0} \int_{x_0}^{x_1} \zeta_1\, dx \; ;$$

then the function

$$\zeta = \zeta_1 - c$$

is continuous and satisfies (15), hence it must satisfy also (16), therefore

[1] The continuity of F_y follows from the continuity (compare the beginning of §4) of y' and from our assumption (B) concerning F; and η' may be supposed continuous, since (9) must hold for all functions η of class D' which vanish at x_0 and x_1, and therefore *a fortiori* for all functions η of class C' which vanish at x_0 and x_1.

[2] *Loc. cit.*, p. 313.

$$\int_{x_0}^{x_1} \zeta N\, dx = \int_{x_0}^{x_1} \zeta_1 (N - \lambda)\, dx = 0 \ , \tag{17}$$

if we denote by λ the constant

$$\lambda = \int_{x_0}^{x_1} N\, dx/(x_1 - x_0) \ .$$

But from (17) it follows by the Fundamental Lemma that[1]

$$N = \lambda \ ,$$

i. e., constant, Q. E. D.

b) Another, more direct, proof has been given by HILBERT[2] in his lectures (summer 1899). He selects arbitrarily four values, $\alpha, \beta, \alpha', \beta'$ satisfying the inequalities

$$x_0 < \alpha < \beta < \alpha' < \beta' < x_1 \ ,$$

and then builds up a function[3] η of class C' which is equal to zero in $(x_0\alpha)$; increases from 0 to a positive value k as x increases from α to β; remains constant, $= k$ in $(\beta\alpha')$; decreases from k to 0 as x increases from α' to β', and finally is equal to zero in $(\beta' x_1)$:

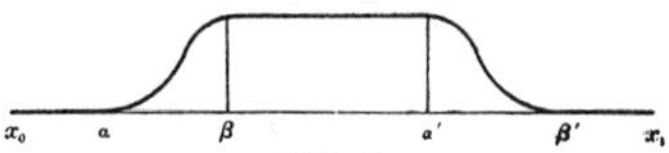

FIG. 3

Substituting this function in (14), we obtain

$$\int_{\alpha}^{\beta} \eta' N\, dx + \int_{\alpha'}^{\beta'} \eta' N\, dx = 0 \ ;$$

η' being positive in the first, and negative in the second, integral we can apply to both the first mean-value theorem[4] which furnishes

$$k \left\{ N\big(\alpha + \theta(\beta - \alpha)\big) - N\big(\alpha' + \theta'(\beta' - \alpha')\big) \right\} = 0 \ ,$$

where $0 < \theta < 1$ and $0 < \theta' < 1$.

Finally, let β and β' approach α and α' respectively; then it follows, since N is continuous, that

[1] This result is a special case of the isoperimetric modification of the Fundamental Lemma, see below chap. vi.

[2] See WHITTEMORE, *Annals of Mathematics* (2), Vol. II (1901), p. 132.

[3] Nothing more than the *existence* of such a function—which is *a priori* clear—is needed for the proof: HILBERT gives a simple example, see WHITTEMORE'S presentation.

[4] Compare E. II A, p. 97; J. I, No. 49; and P., No. 191, IV.

$$N(\alpha) = N(\alpha') \, ,$$

i. e., N is constant in $(x_0 x_1)$.[1]

c) Applying this lemma to (13) we get

$$F_{y'} - \int_{x_0}^{x} F_y \, dx = \lambda \, ,$$

a constant; or

$$F_{y'} = \lambda + \int_{x_0}^{x} F_y \, dx \, . \tag{18}$$

The right-hand side of this equation is differentiable and its derivative is F_y; hence the same must be true of the left-hand side, *i. e.*, the function

$$F_{y'}\big(x, y(x), y'(x)\big) \equiv F_{y'}[x]$$

is differentiable in $(x_0 x_1)$ and

$$\frac{d}{dx} F_{y'} = F_y \, .$$

Thus we find the important corollary to Theorem I that *every solution of our problem with continuous first derivative*—not only those admitting a second derivative—*must satisfy Euler's equation.*

From the fact that $F_{y'}$ is differentiable *follows the existence*[2] *of the second derivative y'' for all values of x for which*

$$F_{y'y'}\big(x, y(x), y'(x)\big) \neq 0 \, . \tag{19}$$

For, if we put

$$y(x+h) - y(x) = k \, , \qquad y'(x+h) - y'(x) = l \, ,$$

then, since the theorem on total differentials[3] is applicable under our assumptions, and since y' is continuous, we have

[1] HILBERT's proof can easily be extended to the case where N, while finite in $(x_0 x_1)$, has a finite number of discontinuities. For, if α and α' are points of continuity, we can always choose β and β' so near to α and α' respectively that N is continuous in $(\alpha\beta)$ and $(\alpha'\beta')$; it follows then as above that $N(\alpha) = N(\alpha')$, *i. e.*, *under the present assumptions N has the* SAME *constant value in all points of continuity.* Hence it follows further that in a point of discontinuity, c:

$$N(c-0) = N(c+0) \, .$$

[2] First pointed out and emphasized by HILBERT in his lectures; see WHITTEMORE, *loc. cit.*

[3] Compare E. II A, pp. 71, 73; J. I, Nos. 86, 127; and P., No. 105.

$$\frac{F_{y'}[x+h]-F_{y'}[x]}{h}=(F_{y'x}+\alpha)+\frac{k}{h}(F_{y'y}+\beta)+\frac{l}{h}(F_{y'y'}+\gamma) \ ,$$

where α, β, γ approach zero as h approaches zero. Hence it follows that if (19) is satisfied,

$$\underset{h=0}{L}\frac{l}{h} \ , \qquad i.\ e.,\ y''$$

exists, and that

$$y''=\frac{F_y-F_{y'x}-y'F_{y'y}}{F_{y'y'}} \ ; \tag{20}$$

moreover, (20) shows that y'' is *continuous* in (x_0x_1).

§7. MISCELLANEOUS REMARKS CONCERNING THE INTEGRATION OF EULER'S EQUATION

a) Euler's differential equation (I) is of the *second order*,[1] as can be seen from the developed form

$$F_y-F_{y'x}-y'F_{y'y}-y''F_{y'y'}=0 \ ; \tag{21}$$

its general solution contains, therefore, two arbitrary constants,

$$y=f(x,\alpha,\beta) \ . \tag{22}$$

The constants α, β have to be determined[2] by the condition that the curve is to pass through the two points A and B:

$$\begin{aligned} y_0&=f(x_0,\alpha,\beta)\\ y_1&=f(x_1,\alpha,\beta) \ . \end{aligned} \tag{23}$$

Every solution of Euler's equation (curve as well as

[1] Unless $F_{y'y'}(x,y,y')$ should be identically zero. In this case Euler's differential equation degenerates either into a finite equation or into the identity : $0=0$ but *never into a differential equation of the first order*. For if $F_{y'y'}\equiv 0$, F must be of the form : $L(x,y)+M(x,y)\,y'$ and (21) reduces to : $L_y-M_x=0$. See also below, under *d*).

If Euler's differential equation degenerates into a finite equation, it is in general impossible to satisfy the initial conditions when the end-points are fixed.

Also in the general case when F contains higher derivatives, Euler's differential equation can never degenerate into a differential equation of odd order; compare FROBENIUS, *Journal für Mathematik*, Vol. LXXXV (1878), p. 206, and HIRSCH, *Mathematische Annalen*, Vol. XLIX (1897), p. 50.

[2] This determination may be impossible; in this case there exists no solution of the problem which is of class C' and lies in the interior of $\mathfrak{R}$.

function) is called, according to KNESER, an *extremal*; there is then a double infinitude of extremals in the plane.

In the *special case when F does not contain x explicitly*, a first integral of (I) can be found immediately.[1] For, if F does not contain x explicitly, we have

$$\frac{d}{dx}(F - y'F_{y'}) = y'(F_y - \frac{d}{dx}F_{y'}) ,$$

and therefore every solution of (I) also satisfies

$$F - y'F_{y'} = \text{const.} \tag{24}$$

Vice versa, every solution of (24), except $y = \text{const.}$, also satisfies (I).

b) EXAMPLE I (see p. 1):

$$F = y\sqrt{1 + y'^2} .$$

Hence

$$F_y = \sqrt{1 + y'^2} , \quad F_{y'} = \frac{yy'}{\sqrt{1 + y'^2}} , \quad F_{y'y'} = \frac{y}{(\sqrt{1 + y'^2})^3} ,$$

and Euler's equation becomes:

$$\sqrt{1 + y'^2} - \frac{d}{dx}\frac{yy'}{\sqrt{1 + y'^2}} = 0 . \tag{I}$$

or, after performing the differentiation,

$$1 + \left(\frac{dy}{dx}\right)^2 - y\frac{d^2y}{dx^2} = 0 .$$

By putting $\dfrac{dy}{dx} = p$, the integration of this differential equation is reduced to two successive quadratures, and the general integral is easily found to be

$$y = \alpha \cosh \frac{x - \beta}{\alpha} .$$

The extremals are therefore catenaries with the x-axis for directrix.

Since F does not contain x, a first integral could have been obtained directly by the corollary (24);

$$F - y'F_{y'} = \frac{y}{\sqrt{1 + y'^2}} = \alpha .$$

[1] Noticed already by EULER, *loc. cit.*, p. 56, in STÄCKEL's translation.

If $\alpha \neq 0$, this leads to the same result as above; for $\alpha = 0$ we obtain $y=0$, which, however, though a solution of (24), is not a solution of Euler's equation.

The general solution of (I) being found, the next step would be so to determine the two constants of integration that the catenary passes through the two given points.[1]

c) Through a given point a, b in the interior of the region[2] $\mathfrak{R}$ *one and but one extremal of class C' can be drawn in a given direction of amplitude*[3] $\omega \left(\neq \pm \frac{\pi}{2}\right)$, *provided that*

$$F_{y'y'}(a, b, b') \neq 0 \,, \tag{25}$$

where $b' = \tan \omega$.

For, if we solve (I) with respect to y'', we obtain for y'' a function of x, y, y' which, according to our assumptions (B), is continuous and has continuous partial derivatives with respect to y, y' at all points of the domain[2] $\mathfrak{T}$ which satisfy (25). Hence the statement follows from CAUCHY's general existence theorem[4] for differential equations.

[1] For this interesting problem we refer to: LINDELÖF-MOIGNO, *loc. cit.*, No. 103; DIENGER, *loc. cit.*, pp. 15–19; TODHUNTER, *Researches in the Calculus of Variations*, pp. 55–58; CARLL, *A Treatise on the Calculus of Variations*, Nos. 60, 61. For SCHWARZ's solution see HANCOCK, "On the Number of Catenaries through Two Fixed Points." *Annals of Mathematics* (1), Vol. X (1896), pp. 159–174.

[2] See §3, *c*).

[3] See §2, *g*).

[4] "Suppose the functions $f_i(x, y_1, y_2, \cdots, y_n)$ and their first partial derivatives with respect to $y_1, y_2, \cdots, y_n$ to be continuous in the domain

$$|x-a| \leqq \rho \,, \quad |y_1 - b_1| \leqq r \,, \cdots, \quad |y_n - b_n| \leqq r \,;$$

let M be the maximum of the absolute values of the functions f_i in this domain, and let l denote the smaller of the two quantities ρ and r/M.

Then there exists one, and but one, system of functions $y_1(x), y_2(x), \cdots, y_n(x)$ which in the interval $|x-a| < l$ are continuous and differentiable, satisfy the differential equations

$$\frac{dy_i}{dx} = f_i(x, y_1, y_2, \cdots, y_n) \,, \quad (i = 1, 2, \cdots, n)$$

and the inequalities $|y_i(x) - b_i| \leqq r$, and take for $x = a$ the values

$$y_1 = b_1, y_2 = b_2, \cdots, y_n = b_n \,."$$

Compare E. II A, pp. 193 and 199, and J. III, Nos. 77–80; also PICARD, *Traité d'Analyse*, Vol. II, chap. xi.

In order to apply the theorem in the present case, replace (21) by the equivalent system.

$$\frac{dy}{dx} = y', \quad \frac{dy'}{dx} = (F_y - F_{y'x} - y'F_{y'y})/F_{y'y'} \,.$$

If, therefore,

$$F_{y'y'}(a, b, p) \neq 0$$

for every finite value of p, one extremal can be drawn from (a, b) in every direction, except the direction of the y-axis.

A problem for which

$$F_{y'y'}(x, y, p) \neq 0$$

at every point (x, y) of the region $\mathfrak{R}$ for every finite value of p, is called, according to HILBERT, a *regular problem.*

d) We consider next the *exceptional case in which Euler's differential equation degenerates into an identity.*

Suppose the left-hand side of (21) vanishes for every system of values x, y, y', y''. Then, since y'' does not occur in the three first terms, it follows that the coefficient of y'' must vanish identically, so that we must have separately

$$F_{y'y'} \equiv 0 \ , \qquad F_y - F_{y'x} - y'F_{y'y} \equiv 0$$

for every x, y, y'. From the first identity it follows that F must be an integral linear function of y', say

$$F(x, y, y') = M(x, y) + N(x, y)\, y' \ .$$

Substituting this value in the second identity, we get

$$M_y = N_x \ ,$$

the well-known integrability condition for the differential expression

$$M\,dx + N\,dy \ .$$

Hence we infer: If M and N and their first partial derivatives are single-valued and continuous in a *simply-connected region* $\mathfrak{S}$ of the x, y-plane, then there exists[1] a function $V(x, y)$, *single-valued and of class* C' in $\mathfrak{S}$ and such that

$$V_x = M \ , \qquad V_y = N \ ,$$

and therefore

$$F(x, y, y') = V_x + V_y y' = \frac{d}{dx} V(x, y) \ .$$

Hence if $\mathfrak{C}: y = f(x)$ be any curve of class C' drawn in $\mathfrak{S}$ between the points $A(x_0, y_0)$ and $B(x_1, y_1)$ our integral $J_{\mathfrak{C}}$ has the value

[1] See PICARD, *Traité d'Analyse*, 4th ed., Vol. I, p. 111.

$$J_{\mathfrak{C}} \equiv \int_{x_0}^{x_1} F(x, y, y')\,dx = V(x_1, y_1) - V(x_0, y_0) \;,$$

and is therefore independent of the path of integration $\mathfrak{C}$ and depends only upon the position of the two end-points.

On account of the continuity of $V(x, y)$, the result remains true for curves $\mathfrak{C}$ with a finite number of corners, as is at once seen by decomposing the integral J in the usual manner.[1]

Vice versa: If the value of the integral $J_{\mathfrak{C}}$ is independent of the path of integration $\mathfrak{C}$ as long as $\mathfrak{C}$ remains in the interior of a region $\mathfrak{S}$ contained in $\mathfrak{R}$, then the function F must be of the form $M(x, y)+N(x, y)y'$, where $M_y = N_x$, for every point (x, y) in the interior of $\mathfrak{S}$ for which $x_0 \leqq x \leqq x_1$.

For let (x_2, y_2) be any inner point of $\mathfrak{S}$ whose abscissa x_2 lies between x_0 and x_1 and y_2', y_2'' two arbitrarily prescribed values; then we can always draw in $\mathfrak{S}$ a curve $\mathfrak{C}: y = f(x)$, of class C'' which passes through (x_0, y_0), (x_1, y_1), (x_2, y_2), and for which $f'(x_2) = y_2'$, $f''(x_2) = y_2''$.

According to our hypothesis, ΔJ must vanish for every admissible variation of $\mathfrak{C}$, whence we infer by the method of §§ 4, 5 that $y = f(x)$ must satisfy Euler's differential equation. The left-hand side of the latter must therefore vanish for the arbitrary system of values $x = x_2$, $y = y_2$, $y' = y_2'$, $y'' = y_2''$, which proves the above statement.

We thus reach the result:[2]

In order that the value of the integral

$$J = \int_{x_0}^{x_1} F(x, y, y')\,dx$$

may be independent of the path of integration it is necessary and sufficient[3] *that Euler's differential equation degenerate into an identity.*

It is clear that in this case there exists no proper[4] extremum of the integral J.

e) We conclude these remarks by considering briefly the inverse problem: *Given a doubly infinite system of curves (functions)*

$$y = f(x, \alpha, \beta) \;,$$

[1] Compare p. 12.

[2] Compare J. III, Nos. 362, 363, and KNESER, *Lehrbuch*, § 51.

[3] Sufficient only if the region $\mathfrak{S}$ is simply-connected.

[4] Compare § 3, *b*).

to determine a function $F(x, y, y')$ so that the given system of curves shall be the extremals for the integral

$$J = \int_{x_0}^{x_1} F(x, y, y')\, dx \ .$$

This problem has always an *infinitude of solutions* which can be obtained by *quadratures*.[1]

For if

$$y'' = G(x, y, y') \tag{26}$$

is the differential equation of the second order[2] whose general solution is the given function $y = f(x, \alpha, \beta,)$ (with α, β as constants of integration), then we must so determine the function $F(x, y, y')$ that (26) becomes identical with Euler's differential equation for F, *i. e.*, according to (21)

$$F_y - F_{y'x} - F_{y'y}\, y' = G F_{y'y'} \ . \tag{27}$$

If we differentiate (27) with respect to y', we get for $M = F_{y'y'}$ a *linear* partial differential equation of the first order, viz.,

$$\frac{\partial M}{\partial x} + y' \frac{\partial M}{\partial y} + G \frac{\partial M}{\partial y'} + G_{y'} M = 0 \ . \tag{28}$$

If

$$\alpha = \phi(x, y, y') \ , \qquad \beta = \psi(x, y, y')$$

is the solution of the two equations

$$y = f(x, \alpha, \beta) \ , \qquad y' = f_x(x, \alpha, \beta)$$

with respect to α and β, and if further

$$\theta(x, \alpha, \beta) = e^{\int G_{y'}\left(x,\, f(x, \alpha, \beta),\, f_x(x, \alpha, \beta)\right) dx} \ ,$$

and

$$\chi(x, y, y') = \theta\left(x, \phi(x, y, y'), \psi(x, y, y')\right) \ ,$$

[1] DARBOUX, *Théorie des surfaces*, Vol. III, Nos. 604, 605. For the analogous problem in the more general case when F contains higher derivatives, compare HIRSCH, *Mathematische Annalen*, Vol. XLIX (1897), p. 49.

[2] Obtained by eliminating α, β between the three equations

$$y = f(x, \alpha, \beta) \ , \quad y' = f_x(x, \alpha, \beta) \ , \quad y'' = f_{xx}(x, \alpha, \beta) \ ;$$

compare, for inst., J. I, No. 166.

the general integral of (28) is found to be, according to the general theory[1] of linear partial differential equations of the first order,

$$M\chi = \Phi\left(\phi(x, y, y'), \psi(x, y, y')\right) ,$$

where Φ is an arbitrary function of ϕ and ψ.

After the function M has been found, F is obtained by two successive quadratures from the differential equation

$$\frac{\partial^2 F}{\partial y'^2} = M(x, y, y') .$$

Finally the two constants of integration λ, μ (which are functions of x and y), introduced by the latter process, must be so determined that F satisfies the original partial differential equation (27) from which (28) was derived by differentiation.

EXAMPLE:[2] To determine all functions F for which the extremals are straight lines

$$y = \alpha x + \beta .$$

The differential equation (26) becomes, in this case,

$$y'' = 0 .$$

Accordingly, we obtain

$$\phi = y' , \qquad \psi = y - xy' , \qquad \chi = \text{const.}$$

Hence

$$M = \Phi(y', y - xy') ,$$

and therefore

$$F = \int_0^{y'} (y' - t)\,\Phi(t, y - xt)\,dt + y'\lambda(x, y) + \mu(x, y) .$$

The condition for λ and μ becomes in this case

$$\frac{\partial\lambda}{\partial x} = \frac{\partial\mu}{\partial y} .$$

The most general expression for λ and μ is therefore

$$\lambda = \frac{\partial\nu}{\partial y} , \qquad \mu = \frac{\partial\nu}{\partial x} ,$$

where ν is an arbitrary function of x and y.

[1] Compare, for inst., J. III, No. 242.

[2] Compare DARBOUX, *loc. cit.*, No. 606.

§8. WEIERSTRASS'S LEMMA AND THE E-FUNCTION

Before proceeding to the consideration of so-called discontinuous solutions, we must derive a lemma, due to WEIERSTRASS,[1] which is of fundamental importance for many investigations in the Calculus of Variations.

Suppose there are given, in the region $\mathfrak{R}$, an extremal $\mathfrak{C}$ of class[2] C'' : $y=f(x)$, and a curve $\tilde{\mathfrak{C}}$ of class C' : $\tilde{y}=\tilde{f}(x)$, meeting $\mathfrak{C}$ at a point[3] $2:(x_2, y_2)$. Besides there is given a point $0:(x_0, y_0)$ on $\mathfrak{C}$, before 2, that is, $x_0<x_2$. Let 3 be that point of $\tilde{\mathfrak{C}}$ whose abscissa is x_2+h, h being a positive infinitesimal, and select arbitrarily a function η of class C' satisfying the conditions $\eta_0\equiv\eta(x_0)=0$, $\eta_2\equiv\eta(x_2)\neq 0$.

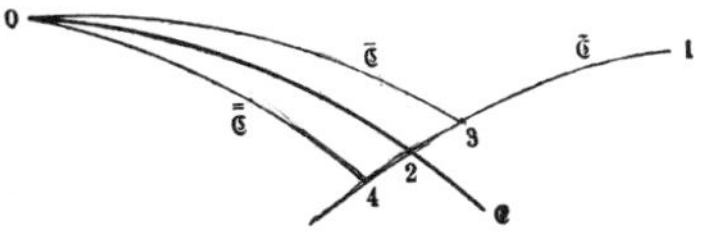

FIG. 4

Then we can so determine ϵ that the curve

$$\bar{\mathfrak{C}}: \qquad \bar{y}=y+\epsilon\eta ,$$

which necessarily passes through the point 0, also passes through the point 3. For this purpose we have to solve the equation

$$f(x_2+h)+\epsilon\,\eta(x_2+h)=\tilde{f}(x_2+h)$$

with respect to ϵ. Since $\tilde{f}(x_2)=f(x_2)$, we have

$$\tilde{f}(x_2+h)-f(x_2+h)=(\tilde{y}_2'-y_2')\,h+h\,(h) ,$$

where $y_2'=f'(x_2)$, $\tilde{y}_2'=\tilde{f}'(x_2)$ and (h) is an infinitesimal for $Lh=0$. Hence we obtain

$$\epsilon=h\left[\frac{\tilde{y}_2'-y_2'}{\eta_2}+(h)\right].$$

It is proposed to compute the difference

$$\Delta J=\bar{J}_{03}-(J_{02}+\tilde{J}_{23}) ,$$

[1] The lemma here given is a modification of the corresponding lemma given by WEIERSTRASS in his lectures (1879) for the case of parameter-representation; see §28.

[2] This assumption must be made on account of the integration by parts which occurs below; compare §4.

[3] For the notation compare §2, *e*).

the integrals J, $\bar{J}$, $\tilde{J}$ being taken along the curves $\mathfrak{E}$, $\bar{\mathfrak{C}}$, $\tilde{\mathfrak{C}}$ respectively, from the point represented by the first index to the point represented by the second.

ΔJ may be written

$$\Delta J = \int_{x_0}^{x_2} (\bar{F} - F)\,dx + \int_{x_2}^{x_2+h} (\bar{F} - \tilde{F})\,dx \;,$$

where $F, \bar{F}, \tilde{F}$ or $F[x], \bar{F}[x], \tilde{F}[x]$ stand for $F(x, y(x), y'(x))$, $F(x, \bar{y}(x), \bar{y}'(x))$, $F(x, \tilde{y}(x), \tilde{y}'(x))$ respectively.

The first integral, treated by the method of §4, becomes, since $\mathfrak{E}$ is an extremal,

$$\begin{aligned}\int_{x_0}^{x_2} (\bar{F} - F)\,dx &= \epsilon\eta_2 F_{y'}[x_2] + \epsilon(\epsilon) \\ &= h\left[(\tilde{y}_2' - y_2')\,F_{y'}[x_2] + (h)\right] .\end{aligned}$$

To the second integral we apply the first mean-value theorem and obtain, on account of the continuity of $F[x]$ and $\tilde{F}[x]$,

$$\int_{x_2}^{x_2+h} (\bar{F} - \tilde{F})\,dx = h\left[F[x_2] - \tilde{F}[x_2] + (h)\right] .$$

Collecting the terms, we reach the result

$$\bar{J}_{03} - (J_{02} + \tilde{J}_{23}) = h\left\{(\tilde{y}_2' - y_2')\,F_{y'}[x_2] + F[x_2] - \tilde{F}[x_2] + (h)\right\} .$$

Similarly let 4 be that point of $\tilde{\mathfrak{C}}$ whose abscissa is $x_2 - h$, and determine ϵ' so that the curve

$$\bar{\bar{\mathfrak{C}}}: \qquad \bar{\bar{y}} = y + \epsilon'\eta$$

passes through 4. Then we obtain by the same process

$$\bar{\bar{J}}_{04} + \tilde{J}_{42} - J_{02} = -h\left\{(\tilde{y}_2' - y_2')\,F_{y'}[x_2] + F[x_2] - \tilde{F}[x_2] + (h)\right\} .$$

If we put for brevity

$$\begin{aligned}F(x, y, \tilde{p}) - F(x, y, p) - (\tilde{p} - p)\,F_{y'}(x, y, p)& \\ = \mathbf{E}(x, y;\; p, \tilde{p})\;,&\qquad (29)\end{aligned}$$

$x, y, p, \tilde{p}$ being considered as four independent variables, the preceding results may be written:

$$\left.\begin{aligned}\bar{J}_{03} - (J_{02} + \tilde{J}_{23}) &= -h\left\{\mathbf{E}(x_2, y_2;\ y_2', \tilde{y}_2') + (h)\right\},\\ \bar{\bar{J}}_{04} + (\tilde{J}_{42} - J_{02}) &= +h\left\{\mathbf{E}(x_2, y_2;\ y_2', \tilde{y}_2') + (h)\right\}.\end{aligned}\right\} \quad (30)$$

We shall refer to these two formulae as *Weierstrass's Lemma.* The function $\mathbf{E}(x, y;\ p, \tilde{p})$ defined by (29) will play a most important part in the sequel; it is called *Weierstrass's* **E**-*function.*[1]

The same results (30) hold if the curves 03 and 04 are of the more general type (5a):

$$\bar{y} = f(x) + \omega(x, \epsilon),$$

where the function $\omega(x, \epsilon)$ vanishes identically for $\epsilon = 0$, has the continuity properties enumerated on p. 18, and satisfies besides the conditions:

$$\omega(x_0, \epsilon) \equiv 0 \text{ for every } \epsilon, \text{ and } \omega_\epsilon(x_2, 0) \neq 0 .$$

For the determination of ϵ we have, in this case, the equation:

$$f(x_2 + h) + \omega(x_2 + h, \epsilon) - \tilde{f}(x_2 + h) = 0 .$$

The resulting value of ϵ is of the same form as above. This follows from the theorem[2] on implicit functions; for if

[1] Compare ZERMELO, *Dissertation*, p. 66.

[2] "If $f(x, y)$ is of class C' in the vicinity of (x_0, y_0) and

$$f(x_0, y_0) = 0, \quad f_y(x_0, y_0) \neq 0,$$

then a positive quantity k being chosen arbitrarily but sufficiently small, another positive quantity h_k can be determined such that for every x in the interval $(x_0 - h_k, x_0 + h_k)$ the equation $f(x, y) = 0$ has one and but one solution y between $y_0 - k$ and $y_0 + k$.

The single-valued function $y = \psi(x)$ thus implicitly defined by the equation: $f(x, y) = 0$, is of class C' in the interval $(x_0 - h_k, x_0 + h_k)$ and

$$\frac{dy}{dx} = -\frac{f_x}{f_y}.$$

Hence

$$y - y_0 = (x - x_0)\left[-\frac{f_x(x_0, y_0)}{f_y(x_0, y_0)} + \alpha\right],$$

where $\underset{x=x_0}{L}\alpha = 0$."

(Compare E. II A, p. 72; J. I, No. 91; P., No. 110).

If $f(x, y)$ is regular in the vicinity of (x_0, y_0), also the function $y = \psi(x)$ is regular in the vicinity of x_0. (Compare E. II B, p. 103, and HARKNESS AND MORLEY, *Introduction to the Theory of Analytic Functions*, No. 156.) For the extension of the theorem to a system of m equations between $m + n$ unknown quantities, see the references just given.

we denote the left-hand side of the preceding equation by $F(h, \epsilon)$, this function is of class C' in the vicinity of $h=0$, $\epsilon=0$; further: $F(0, 0)=0$ and finally $F_\epsilon(0, 0)\neq 0$.

Incidentally we notice here the formula

$$\bar{J}_{03} - J_{02} = \int_{x_0}^{x_2} (\bar{F} - F)\, dx + \int_{x_2}^{x_2+h} \bar{F}\, dx$$
$$= h\left[(\tilde{y}_2' - y_2')\, F_{y'}[x_2] + F[x_2] + (h)\right] ,$$

which holds for negative as well as for positive values of h. Hence it follows that if the arc 02 of the extremal $\mathfrak{E}$ minimizes the integral J, the end-point 0 being fixed while the end-point 2 is movable on the curve $\tilde{\mathfrak{C}}$, then the co-ordinates of the point 2 must satisfy the condition

$$F + (\tilde{y}' - y')\, F_{y'} \Big|^{2} = 0 \; .$$

("*Condition of transversality,*" compare the detailed treatment of the problem with variable end-points in §23.)

§9. DISCONTINUOUS SOLUTIONS

We must now free ourselves from the restriction[1] imposed upon the minimizing curve at the beginning of §4, viz., that y' should be continuous in (x_0x_1), and we propose to determine in this section all those solutions of our problem which present corners—so-called "*discontinuous solutions.*"

a) In the first place, the theorem holds that *also discontinuous solutions must satisfy Euler's differential equation.*

Suppose for simplicity[2] that the minimizing curve $\mathfrak{E}$ has only one corner $C(x_2, y_2)$ between A and B. According to §3, *c*) the integral $J_{\mathfrak{E}}$ is then defined by

$$J_{\mathfrak{E}} = \int_{x_0}^{x_2-0} F(x, y, y')\, dx + \int_{x_2+0}^{x_1} F(x, y, y')\, dx \; , \qquad (31)$$

[1] The assumption that the curve shall lie entirely in the interior of the region $\mathfrak{R}$ will still be retained in this section.

[2] The results can be extended at once to the case of several corners.

the notation indicating that $y'(x_2)$ is defined in the first integral by $y'(x_2-0)$, in the second by $y'(x_2+0)$.

The theorem in question is most easily proved by the method of *partial variation*, which is very useful in many investigations of the Calculus of Variations:

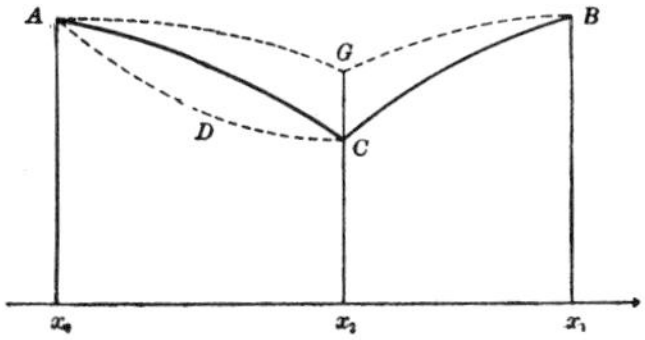

FIG. 5

We consider first such *special*[1] variations ADC of type (5) as leave the arc CB unchanged and vary only AC. To such variations all the conclusions of §§4–6 can be applied, and it follows as before that for the interval (x_0, x_2-0) *Euler's* equation must hold. The same result follows for (x_2+0, x_1) from the consideration of variations which leave AC unchanged; hence it is true for the whole interval (x_0x_1).[2]

b) A discontinuous solution with one corner is therefore composed of two extremals involving in general different constants of integration:

$$y=f(x, \alpha_1, \beta_1) \qquad \text{in } (x_0, x_2-0) \ ,$$
$$y=f(x, \alpha_2, \beta_2) \qquad \text{in } (x_2+0, x_1) \ .$$

For the determination of x_2 and of the constants of integration we have in the first place the initial conditions

$$y_0=f(x_0, \alpha_1, \beta_1) \ ,$$
$$y_1=f(x_1, \alpha_2, \beta_2) \ ;$$

further the condition that y is continuous at x_2:

$$f(x_2, \alpha_1, \beta_1)=f(x_2, \alpha_2, \beta_2) \ ;$$

and finally two further conditions which are furnished by the following theorem due to WEIERSTRASS and ERDMANN:[3]

1 Compare the remark on p. 15, footnote 2).

2 With the same understanding as in (31) concerning the meaning of y' at the corner.

3 WEIERSTRASS, *Lectures* at least as early as 1877; ERDMANN, *Journal für Mathematik*, Vol. LXXXII (1877), p. 21. Another demonstration has been deduced by

THEOREM: *At every corner of a minimizing curve the two limiting values of $F_{y'}$ are equal:*[1]

$$F_{y'}\Big|^{x_2-0} = F_{y'}\Big|^{x_2+0}; \tag{32}$$

and likewise

$$F - y'\,F_{y'}\Big|^{x_2-0} = F - y'\,F_{y'}\Big|^{x_2+0}. \tag{33}$$

To prove (32) consider a variation AGB of type (5) for which the function η is of class C' in $(x_0 x_1)$ and $\eta(x_2) \neq 0$. The integral ΔJ breaks up into two integrals taken between the limits (x_0, x_2-0) and (x_2+0, x_1) respectively. Applying to each of these the methods of §4 we find that also in this case $\delta J = 0$, and further we obtain[2] from (9), since (I) is satisfied:

$$\delta J = \epsilon\eta(x_2)\left(F_{y'}[x_2-0] - F_{y'}[x_2+0]\right),$$

where $F_{y'}[x]$ stands again for $F_{y'}(x, f(x), f'(x))$. Since $\delta J = 0$, (32) is proved.

The proof of (33) follows from Weierstrass's Lemma (30) if we identify the arcs AC and CB of Fig. 5 with the arcs 02 and 21 of Fig. 4, respectively, and consider successively the variations 031 and 04231 of the arc 021. The corresponding values of the total variations ΔJ are given by the two equations (30), the values of y_2', $\bar{y}_2'$ being in the present case

$$y_2' = y'(x_2-0) = \overline{y}_2'\ ; \quad \bar{y}_2' = y'(x_2+0) = \overset{+}{y}_2{}'\ .$$

Hence it follows that for an extremum it is necessary that

WHITTEMORE, *loc. cit.*, from Hilbert's proof of Euler's equation: By means of the extension of the lemma of §6 to discontinuous functions (see p. 25, footnote 1), it can be shown that equation (18) holds with the same value of the constant λ for both segments (x_0, x_2-0) and (x_2+0, x_1). Hence follows Euler's equation as well as equation (32). This method can be applied to discontinuities of a much more complex character and even to the case of an infinitude of points of discontinuity; see WHITTEMORE, *loc. cit.*

[1] For the notation compare §2, *b*).

[2] The integration by parts is legitimate since by the method of §6 the existence of $\frac{d}{dx}F_{y'}$ is established for each of the two segments (x_0, x_2-0) and (x_2+0, x_1).

$$\mathbf{E}(x_2, y_2\ ;\ \overset{-}{y}_2', \overset{+}{y}_2') = 0\ ;$$

and on account of (32) this is equivalent to (33).

c) EXAMPLE[1] III: To minimize the integral

$$J = \int_{x_0}^{x_1} (y' + 1)^2\, y'^2\, dx\ .$$

Here

$$F_y = 0\ , \qquad F_{y'} = 4y'^3 + 6y'^2 + 2y'\ ,$$

$$F - y' F_{y'} = -3y'^4 - 4y'^3 - y'^2\ .$$

Hence a first integral of Euler's differential equation is

$$4y'^3 + 6y'^2 + 2y' = \text{const.}\ ;$$

therefore

$$y = \alpha x + \beta,$$

i.e., the extremals are straight lines, and the line AB joining the two given points is a possible continuous solution.

In order to obtain all discontinuous solutions with one corner, we have to find all solutions p_1, p_2 of the two equations

$$4p_1^3 + 6p_1^2 + 2p_1 = 4p_2^3 + 6p_2^2 + 2p_2\ ,$$

$$-3p_1^4 - 4p_1^3 - p_1^2 = -3p_2^4 - 4p_2^3 - p_2^2\ ,$$

where

$$p_1 = y'(c - 0) \quad \text{and} \quad p_2 = y'(c + 0) \quad \text{and} \quad p_1 \neq p_2\ .$$

Dividing out by $p_1 - p_2$ and putting

$$p_1 + p_2 = u\ , \qquad p_1^2 + p_1 p_2 + p_2^2 = w$$

we get

$$2w + 3u + 1 = 0$$

$$-3u^3 + 6uw + 4w + u = 0\ .$$

These equations have one real solution, $u = -1$, $w = +1$, from which we obtain

$$p_1 = 0\ , \qquad p_2 = -1\ ,$$

or

$$p_1 = -1\ , \qquad p_2 = 0\ .$$

[1] A special case of the example given by ERDMANN, *loc. cit.*, p. 24.

Every discontinuous solution must therefore be composed of straight lines making the angles 0 *or* $3\pi/4$ *with the positive* x-*axis.* If the slope $m=(y_1-y_0)/(x_1-x_0)$ of the line AB lies between 0 and -1, there are indeed two such solutions, AC_1B and AC_2B with one corner and an infinity with $n \gtreqless 2$ corners.

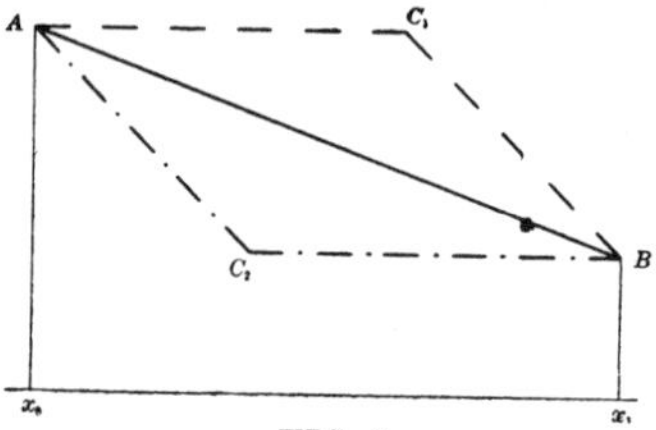

FIG. 6

Since $F=y'^2(y'+1)^2$, these discontinuous solutions furnish for J the value zero and therefore the *absolute minimum*.[1]

d) In many cases the impossibility of discontinuous solutions can be inferred from the following

Corollary:[2] *If* (x_2, y_2) *is a corner of a minimizing curve, then the function*

$$F_{y'y'}(x_2, y_2, p)$$

must vanish for some finite value of p.

For the function

$$\phi(p) = F_{y'}(x_2, y_2, p)$$

is a continuous function of p admitting a finite derivative for all finite values of p; further, if we put

$$y'(x_2-0) = p_1, \quad y'(x_2+0) = p_2 \; ,$$

we have $p_1 \neq p_2$, and, according to (32),

$$\phi(p_1) = \phi(p_2) \; .$$

Hence by Rolle's Theorem the derivative

$$\phi'(p) = F_{y'y'}(x_2, y_2, p)$$

must vanish for some value of p between p_1 and p_2.

If therefore the problem is a "regular problem," *i. e.*, if

$$F_{y'y'}(x, y, p) \neq 0$$

for every point in the interior of $\mathfrak{R}$ and for all finite values

[1] The minimum is, however, "improper" (compare §3, *b*)), because in every neighborhood of AC_1B (or AC_2B) broken lines can be drawn, joining A and B, whose segments have alternately the slopes 0 and -1. For such a curve $\Delta J = 0$.

[2] Compare also WHITTEMORE, *loc. cit.*, p. 136.

of p, we infer that no discontinuous solutions are possible in the interior of $\mathfrak{R}$.

EXAMPLE I (see p. 1): $F=y\sqrt{1+y'^2}$, $\mathfrak{R}$ is the upper half-plane ($y \geqq 0$).[1] Here

$$F_{y'y'}=\frac{y}{\left(\sqrt{1+y'^2}\right)^3}$$

is $\neq 0$ in the interior of $\mathfrak{R}$, and consequently no discontinuous solutions are possible in the interior[2] of $\mathfrak{R}$.

§10. BOUNDARY CONDITIONS

In all the preceding developments it was assumed[3] that the minimizing curve should lie entirely in the interior of the region $\mathfrak{R}$. But there may also exist solutions of the problem as formulated in §3 which have points in common with the boundary of $\mathfrak{R}$. To determine these solutions is the object of the present section.

For this investigation it is convenient to make use of the idea of a *point by point variation* of a curve which played an important part in the earlier history of the Calculus of Variations.

Between the points of the two curves

$$\mathfrak{C}: \qquad y=f(x)\ ,$$

and

$$\overline{\mathfrak{C}}: \qquad y=y+\Delta y$$

we may establish a one-to-one correspondence by letting two points correspond which have the same abscissa x. And we may think of the second curve as being derived from the first by a continuous deformation in which each individual point moves along its ordinate according to some law, for instance, if in

$$y+\alpha\Delta y$$

we let α increase from 0 to 1.

A point of $\mathfrak{C}$ whose abscissa is x', is called a point of *free variation* if $\Delta y(x')$ may take any sufficiently small value; otherwise, a point of *unfree variation.*

For a curve $\mathfrak{C}$ which lies entirely in the interior of $\mathfrak{R}$ all points except the end-points are points of free variation,[4] and this freedom was essential in the conclusions of §§4 and 5.

[1] Compare §1, c).

[2] Compare the next section.

[3] See the beginning of §4.

[4] In our formulation of the problem, §3.

This is not true for a curve which has points in common with the boundary. For simplicity let us suppose that the boundary of $\mathfrak{R}$ contains an arc $\tilde{\mathfrak{C}}$ representable in the form

$$\tilde{y} = \tilde{f}(x) \ ,$$

$\tilde{f}(x)$ being of class C''. In order to fix the ideas suppose that $\mathfrak{R}$ lies above $\tilde{\mathfrak{C}}$. Then if $\mathfrak{C}$ has a point P in common with $\tilde{\mathfrak{C}}$, the variation of P is unfree and restricted by the condition

$$\Delta y \geqq 0 \ . \tag{34}$$

Suppose the minimizing curve 0231 has the *segment 23 in common with the boundary.*

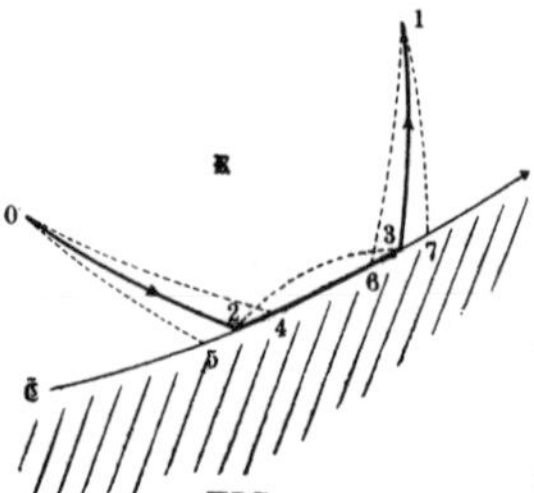

FIG. 7

Then the method of partial variation applied to 02 and to 31 shows that *these two arcs must be extremals.*

Consider next a variation of type (5) which leaves 02 and 31 unchanged and varies only 23. Since $\Delta y = \epsilon\eta$ must be $\geqq 0$, η cannot change sign and if we choose $\eta \geqq 0$ then ϵ *must be taken positive;* hence we can no longer infer from (6) that $\delta J = 0$, but only that

$$\delta J \geqq 0 \ . \tag{35}$$

After the integration by parts of §4 we obtain therefore

$$\int_{x_2}^{x_3} \eta \left(F_y - \frac{d}{dx} F_{y'} \right) dx \geqq 0$$

for all functions η of class D' which vanish at x_2 and x_3 and satisfy besides the condition

$$\eta \geqq 0 \ .$$

The lemma of §5, slightly modified, leads in the present case to the

[1] Moreover at the end-points 2 and 3 the following condition must be satisfied:

$$\mathbf{E}(x_2, y_2;\ y_2', \tilde{y}_2') = 0; \qquad \mathbf{E}(x_3, y_3;\ y_3', \tilde{y}_3') = 0 \ .$$

The proof follows easily from Weierstrass's Lemma (see Fig. 7). Compare also the treatment of the problem in parameter-representation, §29. The question of *sufficient conditions for one-sided variations* has recently been considered by BLISS in a paper read before the Chicago section of the American Mathematical Society. He finds that for a so-called *regular problem* (§7, *c*) the arc 23 of the curve $\mathfrak{T}$ furnishes a

Theorem:[1] *If the minimizing curve has a segment*[1] *23 in common with the boundary of* $\mathfrak{R}$, *then along this segment the following condition must be satisfied:*

$$F_y - \frac{d}{dx} F_{y'} \geqq 0 \ , \quad \text{if } \mathfrak{R} \text{ lies above } 23 \ , \tag{36a}$$

$$F_y - \frac{d}{dx} F_{y'} \leqq 0 \ , \quad \text{if } \mathfrak{R} \text{ lies below } 23 \ . \tag{36b}$$

smaller value for the integral J than any other curve of class D' joining the two points 2 and 3, lying in a certain neighborhood of the arc 23 and *satisfying the condition* $\Delta y \geqq 0$, provided that the condition

$$F_y - \frac{d}{dx} F_{y'} > 0$$

is fulfilled along the arc 23.

The proof is based upon the construction of a "field" (see §§19, 20, 21) of extremals each one of which is tangent to the curve $\mathfrak{T}$ and lies entirely on one side of $\mathfrak{T}$.

[1] Of the properties specified above.

CHAPTER II

THE SECOND VARIATION

§11. LEGENDRE'S CONDITION

THE integration of Euler's differential equation and the subsequent determination of the constants of integration[1] yield in general a certain number[2] of curves $\mathfrak{C}$ as the only possible solutions of our problem; that is, if there exist at all curves which minimize the integral J, they must be contained among these curves.

We have now to examine each one of these curves separately and to decide whether it actually furnishes a minimum or not.

We confine ourselves in this investigation to curves which lie entirely in the interior of the region $\mathfrak{R}$ and have no corners.

a) Generalities concerning the second variation.

We suppose then we have found an extremal

$$\mathfrak{C}_0 : \qquad y = f_0(x) , \qquad x_0 \leqq x \leqq x_1 \tag{1}$$

of class C' which passes through the two points A and B, and which lies entirely in the interior of the region $\mathfrak{R}$.

Then we replace, as in §4, the curve $\mathfrak{C}_0$ by a neighboring curve

$$\bar{y} = y + \omega$$

and apply to the increment ΔJ Taylor's formula,[3] stopping,

[1] By the initial conditions (23), the corner conditions (32) and (33), and the boundary conditions.

[2] The number may be infinite (see Example III, p. 40); but it may also be impossible so to determine the constants as to satisfy the conditions imposed upon them; this happens, for instance, in Example I for certain positions of the two given points; see the references given on p. 28.

[3] If F is an analytic function, regular in the domain $\mathfrak{T}$, expansion into an infinite series may be used instead.

however, at the terms of the third order. If we put for brevity

$$\left.\begin{aligned} F_{yy}\left(x, f_0(x), f_0'(x)\right) &= P \\ F_{yy'}\left(x, f_0(x), f_0'(x)\right) &= Q \\ F_{y'y'}\left(x, f_0(x), f_0'(x)\right) &= R \end{aligned}\right\} \tag{2}$$

and remember that $\delta J = 0$, since $\mathfrak{E}_0$ is an extremal, we obtain

$$\Delta J = \tfrac{1}{2}\int_{x_0}^{x_1} (P\omega^2 + 2Q\omega\omega' + R\omega'^2)\,dx + \int_{x_0}^{x_1} (\omega,\, \omega')_3\,dx\ , \tag{3}$$

$(\omega, \omega')_3$ being a homogeneous function of dimension three of ω, ω'.

Considering again special variations of the type $\omega = \epsilon\eta$ and reasoning as in §4, we obtain

$$\Delta J = \epsilon^2\left[\tfrac{1}{2}\int_{x_0}^{x_1} (P\eta^2 + 2Q\eta\eta' + R\eta'^2)\,dx + (\epsilon)\right]\ , \tag{4}$$

where (ϵ) is again an infinitesimal.

Hence we infer the theorem:

For a minimum (maximum) it is necessary that the second variation be positive (negative) or zero:

$$\delta^2 J \geqq 0 \qquad (\leqq 0) \tag{5}$$

for all functions η of class D' which vanish at x_0 and x_1.

For according to the definition given in §4, *c*),

$$\delta^2 J = \epsilon^2 \int_{x_0}^{x_1} (P\eta^2 + 2Q\eta\eta' + R\eta'^2)\,dx\ . \tag{5a}$$

The same result can also be obtained by the method of differentiation with respect to ϵ, explained in §4, *b*); see p. 16, footnote 2.

From our assumptions concerning the functions $F(x, y, p)$ and $f_0(x)$ it follows[1] that the three functions P, Q, R are continuous in the interval (x_0x_1). We suppose in the sequel that they are not all three identically zero in (x_0x_1).

[1] Compare J. I, No. 60, and P., No. 99.

b) Legendre's condition.

For the discussion of the sign of the second variation, LEGENDRE[1] uses the following artifice: He adds to the second variation the integral

$$\epsilon^2 \int_{x_0}^{x_1} (2\eta\eta' w + \eta^2 w')\, dx \ ,$$

where w is an arbitrary function of x of class C' in (x_0x_1). This integral is equal to zero;[2] for it is equal to

$$\epsilon^2 \int_{x_0}^{x_1} \frac{d}{dx}(\eta^2 w)\, dx = \epsilon^2 \Big[\eta^2 w\Big]_{x_0}^{x_1} ,$$

and η vanishes at x_0 and x_1.

He thus obtains $\delta^2 J$ in the form

$$\delta^2 J = \epsilon^2 \int_{x_0}^{x_1} \left[(P + w')\eta^2 + 2(Q + w)\eta\eta' + R\eta'^2\right] dx \ .$$

And now he determines the arbitrary function w by the condition that the discriminant of the quadratic form in η, η' under the integral shall vanish, *i. e.*,

$$(Q + w)^2 - R(P + w') = 0 \ . \qquad (6)$$

This reduces $\delta^2 J$ to the form

$$\delta^2 J = \epsilon^2 \int_{x_0}^{x_1} R\left(\eta' + \frac{Q + w}{R}\eta\right)^2 dx \ , \qquad (7)$$

from which he infers that R must not change sign in (x_0x_1) and that $\delta^2 J$ has then always the same sign as R.

These conclusions are, however, open to objections. For, as LAGRANGE[3] had already remarked, Legendre's transformation tacitly presupposes that the differential equation

[1] LEGENDRE: "Mémoire sur la manière de distinguer les maxima des minima dans le calcul des variations," *Mémoires de l'Académie des Sciences*, 1786; in STÄCKEL's translation in OSTWALD's *Klassiker der exacten Wissenschaften*, No. 47, p. 59.

[2] This holds true also when η has discontinuities of the kind which we have admitted (§3, *c*)); compare p. 12, footnote 5), and remember that η and w are continuous in (x_0x_1).

[3] In 1797; see *Oeuvres*, Vol. IX, p. 303.

(6) has an integral which is finite and continuous in the interval (x_0x_1), and that R does not vanish in (x_0x_1).

Nevertheless, by a slight modification[1] of the reasoning, the first part of Legendre's conclusion can be rigorously proved, *i. e.*, the

FUNDAMENTAL THEOREM II: *For a minimum (maximum) it is necessary that*

$$R(x) \equiv F_{y'y'}\left(x, f_0(x), f_0'(x)\right) \geqq 0 \; (\leqq 0) \quad \textit{in} \quad (x_0x_1) \,. \tag{II}$$

For, suppose $R(c) < 0$ for some value c in (x_0x_1); then we can assign a subinterval $(\xi_0\xi_1)$ of (x_0x_1) for which the following two conditions are simultaneously fulfilled:

1. $R(x) < 0$ throughout $(\xi_0\xi_1)$;
2. There exists a particular integral w of (6) which is of class C' in $(\xi_0\xi_1)$.

For, since $R(x)$ is continuous in (x_0x_1) and $R(c) < 0$, we can determine a vicinity $(c-\delta,\ c+\delta)$ of c in which $R(x) < 0$. Hence it follows that if we write the differential equation (6) in the form

$$\frac{dw}{dx} = -P + \frac{(Q+w)^2}{R}\,, \tag{6a}$$

the right-hand side, considered as a function of x and w, is continuous and has a continuous partial derivative with respect to w in the vicinity of the point $x=c$, $w=w_0$, w_0 being an arbitrary initial value for w.

Hence there exists, according to CAUCHY'S existence theorem,[2] an integral of (6) which takes for $x=c$ the value $w=w_0$, and which is of class C' in a certain vicinity $(c-\delta',\ c+\delta')$ of c. The interval $(\xi_0\xi_1)$ in question is the smaller of the two intervals $(c-\delta,\ c+\delta)$ and $(c-\delta',\ c+\delta')$.

This point being established, we choose for η a function which is identically zero outside of $(\xi_0\xi_1)$, and equal to

[1] The proof in the text follows WEIERSTRASS'S exposition, *Lectures*, 1879.

[2] Compare p. 28, footnote 4.

$(x-\xi_0)(x-\xi_1)$ in $(\xi_0\xi_1)$. The function η thus defined furnishes an admissible variation of the curve $\mathfrak{E}_0$, since it is of class D' in (x_0x_1), and vanishes at x_0 and x_1.

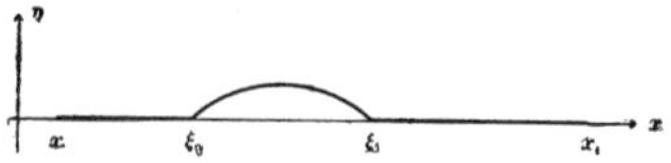

FIG. 8

For this particular function η, $\delta^2 J$ becomes

$$\delta^2 J = \epsilon^2 \int_{\xi_0}^{\xi_1} (P\eta^2 + 2Q\eta\eta' + R\eta'^2)\,dx \ .$$

To this integral Legendre's transformation is applicable. Accordingly

$$\delta^2 J = \epsilon^2 \int_{\xi_0}^{\xi_1} R\left(\eta' + \frac{Q+w}{R}\eta\right)^2 dx \ .$$

The function $\eta' + \dfrac{Q+w}{R}\eta$ is certainly not identically zero throughout $(\xi_0\xi_1)$; for it is different from zero for $x = \xi_0$ and $x = \xi_1$.

Hence if $R(c)$ were negative, a variation of $\mathfrak{E}_0$ could be found for which $\delta^2 J < 0$, which is impossible if $\mathfrak{E}_0$ minimizes the integral J. Therefore $R(x) \geqq 0$ in (x_0x_1), Q. E. D.

Leaving aside the exceptional case[1] in which $R(x)$ has zeros in the interval (x_0x_1), we assume in the sequel that for the extremal $\mathfrak{E}_0$ the condition

$$R > 0 \qquad \text{in } (x_0x_1) \tag{II'}$$

is fulfilled.

A consequence of this assumption is that not only $f_0'(x)$ but also $f_0''(x)$ is continuous in (x_0x_1), as follows immediately from equation (20) at the end of §6. Hence we infer that not only the functions P, Q, R themselves but also their first derivatives are continuous in (x_0x_1).

EXAMPLE[2] I (see p. 27): $F = y\sqrt{1+y'^2}$; hence

[1] An example of this exceptional case is considered by ERDMANN, *Zeitschrift für Mathematik und Physik*, Vol. XXIII (1878), p. 369, viz.,

$$F = y^2 \cos^2 x \qquad \text{and} \qquad x_0 < \frac{\pi}{2} < x_1 \ .$$

[2] All the square roots are to be taken positive, see p. 2, footnote 1.

$$F_{yy}=0\ , \qquad F_{yy'}=\frac{y'}{\sqrt{1+y'^2}}\ , \qquad F_{y'y'}=\frac{y}{\left(\sqrt{1+y'^2}\right)^3}\ .$$

Further

$$\mathfrak{E}_0: \qquad y=a_0\cosh\frac{x-\beta_0}{a_0}\ ,$$

hence

$$P=0\ , \qquad Q=\tanh\frac{x-\beta_0}{a_0}\ , \qquad R=a_0/\cosh^2\frac{x-\beta_0}{a_0}\ .$$

Since we suppose $y>0$, it follows that $a_0>0$ and therefore $R>0$ for every x.

c) *Jacobi's form of Legendre's differential equation.*

We have now to examine the second part of Legendre's conclusion, viz., that, if $R>0$ throughout (x_0x_1), then $\delta^2J\geqq 0$ for all admissible functions η.

The conclusion is correct, as follows immediately from the preceding developments, whenever there exists an integral of the differential equation (6) which is finite and continuous[1] throughout (x_0x_1); it is wrong, as will be seen in §16, if no such integral exists.

It is therefore necessary to enter into a discussion of the differential equation (6). For this purpose JACOBI[2] reduces the differential equation (6) to a homogeneous linear differential equation of the second order by the substitution[3]

$$w=-Q-R\frac{u'}{u}\ , \tag{8}$$

which transforms (6) into

$$(P-Q')\,u-\frac{d}{dx}(Ru')=0\ . \tag{9}$$

We shall refer to this differential equation as *Jacobi's differential equation* and shall denote its left-hand side by $\Psi(u)$:

[1] Since $R\neq 0$, the continuity of w implies the continuity of w, compare (6a).

[2] "Zur Theorie der Variations-Rechnung und der Differentialgleichungen," *Journal für Mathematik*, Vol. XVII (1837), p. 68; also *Ostwald's Klassiker*, etc., No. 47, p. 87.

[3] Notice that also the derivatives of Q, R exist and are continuous, as shown above.

$$\Psi(u) \equiv \left(P - \frac{dQ}{dx}\right)u - \frac{d}{dx}\left(R\frac{du}{dx}\right) . \tag{10}$$

If we write (9) in the form

$$\frac{d^2u}{dx^2} + \frac{R'}{R}\frac{du}{dx} + \frac{Q' - P}{R}u = 0 , \tag{9a}$$

the coefficients are continuous in (x_0x_1). Hence it follows, according to the general existence theorem[1] on linear differential equations, that every integral of (10) is continuous and admits continuous first and second derivatives in (x_0x_1).

Hence we can infer that *if the condition: $R > 0$ in (x_0x_1) is satisfied and if the differential equation* (9) *has an integral u which is different from zero throughout* (x_0x_1), *then $\delta^2J > 0$ for every admissible function η not identically zero.*

For if u is such an integral, then (8) furnishes an integral w of (6) of class C′ in (x_0x_1), and therefore $\delta^2J \geqq 0$. In order to show that the equality sign must be excluded, we introduce u instead of w in (7), and obtain

$$\delta^2J = \epsilon^2 \int_{x_0}^{x_1} \frac{R(\eta' u - \eta u')^2}{u^2} dx . \tag{11}$$

This shows that δ^2J can be equal to zero only when $\eta' u - \eta u' \equiv 0$ throughout (x_0x_1), *i. e.*, when $\eta = \text{Const. } u$, which is impossible since η vanishes at x_0 and x_1, and u does not.

If, on the contrary, every integral of (9) vanishes at least at one point of (x_0x_1), Legendre's tranformation is not applicable to the whole interval. We shall see (in §16) that in this case δ^2J can, in general, be made negative.

[1] Compare E. II A, p. 194, and PICARD, *Traité d'Analyse*, 3d ed., Vol. III, p. 95. If F and consequently also P, Q, R are analytic functions, the existence theorems for analytic differential equations may be used instead. For linear differential equations in particular, see SCHLESINGER, *Handbuch der Theorie der linearen Differentialgleichungen*, Vol. I, p. 21.

§12. JACOBI'S TRANSFORMATION OF THE SECOND VARIATION

The proof of the statement made at the end of the preceding section is based upon a second transformation of $\delta^2 J$ due to JACOBI.[1]

a) Let $(\xi_0\xi_1)$ be either the interval (x_0x_1) itself or a subinterval of (x_0x_1), and let η be identically zero outside of $(\xi_0\xi_1)$, and in $(\xi_0\xi_1)$ equal to some function of class C'' which vanishes at ξ_0 and ξ_1.

Then if we denote by 2Ω the quadratic form of η, η':

$$2\Omega = P\eta^2 + 2Q\eta\eta' + R\eta'^2 ,$$

and apply Euler's theorem on homogeneous functions, we may write $\delta^2 J$ in the form

$$\delta^2 J = \epsilon^2 \int_{\xi_0}^{\xi_1} \left(\eta \frac{\partial \Omega}{\partial \eta} + \eta' \frac{\partial \Omega}{\partial \eta'} \right) dx .$$

The second term can be integrated by parts since η'' is continuous, and we obtain

$$\delta^2 J = \epsilon^2 \left\{ \left[\eta \frac{\partial \Omega}{\partial \eta'} \right]_{\xi_0}^{\xi_1} + \int_{\xi_0}^{\xi_1} \eta \left(\frac{\partial \Omega}{\partial \eta} - \frac{d}{dx}\frac{\partial \Omega}{\partial \eta'} \right) dx \right\} .$$

[1] *Journal für Mathematik*, Vol. XVII (1837), p. 68. JACOBI derives (12) as well as the integration of (9) from the remark that $\delta^2 J = \delta(\delta J)$, hence

$$\delta^2 J = \epsilon \left\{ \left[\eta \, \delta F_{y'} \right]_{x_0}^{x_1} + \int_{x_0}^{x_1} \eta \, \delta M \, dx \right\} ,$$

where

$$M = F_y - \frac{d}{dx} F_{y'} .$$

But

$$\delta M = \Psi(\delta y) = \epsilon \Psi(\eta) .$$

Jacobi's paper, which is not confined to the simple case which we are here considering, but which also treats the case in which the function F contains higher derivatives of y of any order, marks a turning point in the history of the Calculus of Variations. It gives, however, only very short indications concerning the proofs; the details of the proofs have been supplied in a series of articles by DELAUNAY, SPITZER, HESSE and others (see the list given by PASCAL, *loc. cit.*, p. 63). Among these commentaries on Jacobi's paper, the most complete is that by HESSE (*Journal für Mathematik*, Vol. LIV (1857), p. 255), whose presentation we follow in this section. JACOBI, *Werke*, Vol. IV, pp. 39–41. HESSE, *Werke*, p. 446.

Jacobi's results have been extended to the most general problem involving simple definite integrals by CLEBSCH and A. MAYER (see the references given in PASCAL, *loc. cit.*, pp. 64, 65, and C. JORDAN, *Cours d'Analyse*, Vol. III, Nos. 373–94).

But η vanishes at ξ_0 and ξ_1, and

$$\frac{\partial \Omega}{\partial \eta} - \frac{d}{dx}\frac{\partial \Omega}{\partial \eta'} = (P - Q')\eta - \frac{d}{dx}(R\eta') = \Psi(\eta) \ .$$

Hence we obtain JACOBI'S *expression for the second variation:*

$$\delta^2 J = \epsilon^2 \int_{\xi_0}^{\xi_1} \eta \Psi(\eta)\, dx \ , \tag{12}$$

which leads at once to the following result:

If there exists an integral u of the differential equation (9) which vanishes at two points ξ_0 and ξ_1 of $(x_0 x_1)$, we can make[1] $\delta^2 J = 0$, viz., by choosing

$$\eta = \begin{cases} u \text{ in } (\xi_0 \xi_1) \ , \\ 0 \text{ outside of } (\xi_0 \xi_1) \ . \end{cases}$$

b) In the sequel we shall need an *extension of formula (12) to the case when η is of class D''.* Let $c_1, c_2, \cdots, c_n$ be the points of discontinuity of η' or η''. Then the integral for $\delta^2 J$ must be broken up into a sum of integrals from ξ_0 to c_1, from c_1 to c_2, etc., before the integration by parts is applied. Hence we obtain in this case

$$\delta^2 J = \epsilon^2 \left\{ \sum_{\nu=1}^{n} \left[\eta \frac{\partial \Omega}{\partial \eta'} \right]_{c_\nu + 0}^{c_\nu - 0} + \int_{\xi_0}^{\xi_1} \eta \Psi(\eta)\, dx \right\} .$$

or, if we substitute for $\frac{\partial \Omega}{\partial \eta'}$ its value and remember that η, Q, R are continuous at $c_1, c_2, \cdots, c_n$:

$$\delta^2 J = \epsilon^2 \left\{ \sum_{\nu=1}^{n} \eta(c_\nu) R(c_\nu) \left[\eta'(c_\nu - 0) - \eta'(c_\nu + 0) \right] + \int_{\xi_0}^{\xi_1} \eta \Psi(\eta)\, dx \right\} . \tag{12a}$$

c) From (12) *a second proof*[2] *of (11)* can be derived; this proof is based upon the following property of the differen-

[1] It will be seen later on that it follows from this result that, in general, there can be no extremum in this case, see §§14 and 16.

[2] Due to JACOBI, see the references on p. 51, footnote 1, in particular to HESSE.

tial operator Ψ: If u and v are any two functions of class C'', then

$$u\Psi(v) - v\Psi(u) = -\frac{d}{dx}R(uv' - u'v) \, . \qquad (13)$$

Hence if u satisfies the differential equation

$$\Psi(u) = 0 \, ,$$

we get

$$u\Psi(v) = -\frac{d}{dx}R(uv' - u'v) \, ,$$

and if we put

$$v = pu \, ,$$

p being any function of class C'', and multiply by p, we obtain

$$(pu)\Psi(pu) = -p\frac{d}{dx}(Rp'u^2)$$

$$= -\frac{d}{dx}(Rpp'u^2) + R(p'u)^2 \, . \qquad (14)$$

But since

$$Pv^2 + 2Qvv' + Rv'^2 = v\Psi(v) + \frac{d}{dx}v(Qv + Rv')$$

we obtain from (14):

$$P(pu)^2 + 2Q(pu)\frac{d(pu)}{dx} + R\left(\frac{d(pu)}{dx}\right)^2$$

$$= R(p'u)^2 + \frac{d}{dx}\left(p^2u(Qu + Ru')\right) \, . \qquad (15)$$

Now *suppose moreover that* u *is different from zero throughout* $(\xi_0\xi_1)$. Then we may substitute in (15) for the arbitrary function p the quotient

$$p = \frac{\eta}{u} \, ;$$

and since η vanishes at ξ_0 and ξ_1, also p will vanish at

ξ_0 and ξ_1. Hence, on integrating (15) between the limits ξ_0 and ξ_1, and substituting for p its value, we obtain[1]

$$\delta^2 J = \epsilon^2 \int_{\xi_0}^{\xi_1} \frac{R(\eta' u - \eta u')^2}{u^2} dx \ . \tag{11a}$$

§13. JACOBI'S THEOREM

By the developments of the last two sections, the decision regarding the sign of the second variation is reduced to the discussion of Jacobi's differential equation (9). It is therefore a theorem of fundamental importance, discovered by JACOBI[2] in 1837, that the general solution of the differential equation $\Psi(u)=0$ can be obtained by mere processes of differentiation, as soon as the general solution of Euler's differential equation is known.

a) Assumptions[3] *concerning the general solution* $f(x, \alpha, \beta)$ *of Euler's differential equation:*

We suppose for this investigation that the extremal $\mathfrak{E}_0$ is derived from the general solution by giving the constants α, β the special values α_0, β_0, so that

$$f_0(x) = f(x, \alpha_0, \beta_0) \ .$$

Further, we suppose that the function $f(x, \alpha, \beta)$, its first

[1] Notice that in the present proof we have to suppose η to be of class C'' in $(\xi_0\xi_1)$. It can, however, be easily proved that the result is true also for functions η of class C' and even D', in accordance with the results of §11, c). This follows from the fact that p'' does not occur in the identity (15) and that $p^2u(Qu+Ru')$ is continuous even at the points of discontinuity of η' or η''.

[2] See the reference on p. 51, footnote.

[3] If the interval (x_0x_1) is sufficiently small, these assumptions are a consequence of our previous assumptions concerning the function F (p. 12), the extremal $\mathfrak{E}_0$ (p. 44) and the function R (p. 48). This follows from the theorems concerning the dependence of the general solution of a system of differential equations upon the constants of integration; compare PAINLEVÉ in E. II A, pp. 195 and 200, and the references there given to PICARD, BENDIXSON, PEANO, NICOLETTI, and V. ESCHERICH; also NICOLETTI, *Atti della R. Acc. dei Lincei Rendiconti*, 1895, p. 816.

For the case when F is an analytic function, compare E. II A, p. 202, and KNESER, *Lehrbuch*, §27.

For certain special investigations concerning the "conjugate points," the additional assumption is necessary that also $f_{\alpha\alpha}, f_{\alpha\beta}, f_{\beta\beta}$ exist and are continuous in $\mathfrak{A}$; compare p. 59, footnote 1, and p. 62, footnote 4.

partial derivatives and the cross-derivatives $f_{x\alpha}$, $f_{x\beta}$ are continuous, and that f_{xx} exists in a certain domain

$$\mathfrak{A}: \qquad X_0 \leqq x \leqq X_1 \,, \qquad |\alpha - \alpha_0| \leqq d \,, \qquad |\beta - \beta_0| \leqq d \,,$$

where $X_0 < x_0$, $X_1 > x_1$ and d is a positive quantity.

From these assumptions, together with our previous assumptions concerning the function F, the assumption that $\mathfrak{E}_0$ lies in the interior of the region $\mathfrak{R}$ and the assumption that $R(x) > 0$ in (x_0x_1) it follows:

1. That[1] also the partial derivatives $f_{\alpha x}$, $f_{\beta x}$ exist, are continuous and equal to $f_{x\alpha}$, $f_{x\beta}$ respectively, throughout $\mathfrak{A}$;

2. That if we replace in the first and second partial derivatives of F the arguments y, y' by $f(x, \alpha, \beta)$, $f_x(x, \alpha, \beta)$, these partial derivatives are changed into functions of x, α, β which are continuous and have continuous first partial derivatives with respect to α and β;

3. That[2]

$$F_{y'y'}\big(x, f(x, \alpha, \beta), f_x(x, \alpha, \beta)\big) > 0 \,, \tag{16}$$

the last two statements being true throughout the domain $\mathfrak{A}$ provided that the quantity d and the differences $x_0 - X_0$, $X_1 - x_1$ be taken sufficiently small;

4. The quantities d, $x_0 - X_0$, $X_1 - x_1$ being so selected, it follows further from equation (20) in §6 that also the partial derivatives f_{xx}, $f_{xx\alpha}$, $f_{xx\beta}$ exist and are continuous in $\mathfrak{A}$.

b) The general integral of Jacobi's differential equation (9) can now be obtained according to JACOBI (*loc. cit.*) as follows:

If we substitute in Euler's differential equation for y the general integral $f(x, \alpha, \beta)$ we obtain

[1] Compare E. II A, p. 73, and STOLZ, *Grundzüge der Differential- und Integralrechnung*, Vol. I, p. 150.

[2] Since $R(x)$ has a positive minimum value in (x_0x_1) and $F_{y'y'}\big(x, f(x, \alpha, \beta), f_x(x, \alpha, \beta)\big)$ is uniformly continuous in $\mathfrak{A}$.

$$F_y\big(x, f(x, \alpha, \beta), f_x(x, \alpha, \beta)\big) \\ - \frac{d}{dx} F_{y'}\big(x, f(x, \alpha, \beta), f_x(x, \alpha, \beta)\big) = 0 \ ,$$

an identity which is satisfied for all values of x, α, β in the domain $\mathfrak{A}$ and which may therefore be differentiated with respect to α or β. On account of the preceding assumptions, the order of differentiation with respect to x and α (or β) may be reversed[1] and we obtain

$$\begin{aligned} \left(F_{yy} - \frac{d}{dx} F_{y'y}\right) f_\alpha - \frac{d}{dx}\left(F_{y'y'} f'_\alpha\right) = 0 \ , \\ \left(F_{yy} - \frac{d}{dx} F_{y'y}\right) f_\beta - \frac{d}{dx}\left(F_{y'y'} f'_\beta\right) = 0 \ , \end{aligned} \tag{17}$$

where the accents denote again differentiation with respect to x.

If we give in (17) to α, β the particular values $\alpha = \alpha_0$, $\beta = \beta_0$ and remember the definition of P, Q, R in §11 equation (2), we obtain

JACOBI'S *Theorem: If*

$$y = f(x, \alpha, \beta)$$

is the general solution of Euler's differential equation, then the differential equation

$$\Psi(u) \equiv (P - Q')u - \frac{d}{dx}(Ru') = 0$$

admits the two particular integrals

$$\begin{aligned} r_1 = f_\alpha(x, \alpha_0, \beta_0) \\ r_2 = f_\beta(x, \alpha_0, \beta_0) \ . \end{aligned} \tag{18}$$

Corollary:[2] *The two particular integrals r_1 and r_2 are, in general, linearly independent.*

For, in order that r_1 and r_2 may be linearly independent,

[1] From the existence and continuity of $\frac{d}{dx}(F_{y'y'} f_{\alpha x})$ and $\frac{d}{dx} F_{y'y'}$ follows the existence and continuity of $f_{\alpha xx}$ on account of (16).

[2] See PASCAL, *loc. cit.*, p. 75.

it is necessary and sufficient that their "Wronskian determinant"[1]

$$D(x) = \begin{vmatrix} r_1(x) & r_2(x) \\ r_1'(x) & r_2'(x) \end{vmatrix}$$

be not identically zero.

On the other hand, since $f(x, \alpha, \beta)$ is supposed to be the general solution of Euler's differential equation, it must be possible so to determine α and β that y and y' take arbitrarily prescribed values y_2 and y_2' for a given non-singular value of x, say x_2.

The two functions $f(x_2, \alpha, \beta)$ and $f_x(x_2, \alpha, \beta)$ of α, β must therefore be independent, and consequently[2] their Jacobian

$$\frac{\partial(f, f_x)}{\partial(\alpha, \beta)} = \begin{vmatrix} f_\alpha & f_\beta \\ f_{x\alpha} & f_{x\beta} \end{vmatrix}$$

cannot be identically zero for all values of α, β. But for $\alpha = \alpha_0$, $\beta = \beta_0$, this Jacobian is identical with the determinant $D(x)$, since $f_{\alpha x} = f_{x\alpha}$, $f_{\beta x} = f_{x\beta}$, and therefore r_1 and r_2 are linearly independent, except, possibly, for singular systems of values α_0, β_0, *i. e.*, for singular positions of the two given points A and B.

We exclude in the sequel such exceptional cases and assume that r_1 and r_2 are linearly independent. Then *the general integral of Jacobi's differential equation is*

$$u = C_1 r_1 + C_2 r_2 \,, \tag{19}$$

C_1, C_2 being two arbitrary constants.

§14. JACOBI'S CRITERION

By Jacobi's theorem the further discussion of the sign of $\delta^2 J$ is reduced to the question: Under what conditions is it possible so to determine the two constants C_1, C_2 that the function $u = C_1 r_1 + C_2 r_2$ shall not vanish in $(x_0 x_1)$?

[1] Compare E. II A, p. 261, and J. III, No. 122.

[2] Compare P., No. 122, IV and J. I, No. 94.

In order to answer this question, we construct the expression[1]

$$\Delta(x, x_0) = r_1(x) r_2(x_0) - r_2(x) r_1(x_0) \ ; \tag{20}$$

it is a particular integral of (9) and vanishes for $x = x_0$; if it vanishes at all for values of $x > x_0$, let x_0' be the zero next greater than x_0, so that

$$\begin{aligned} &\Delta(x_0, \ x_0) = 0 \ , \\ &\Delta(x \ , \ x_0) \neq 0 \quad \text{for} \quad x_0 < x < x_0' \ , \\ &\Delta(x_0', \ x_0) = 0 \ . \end{aligned} \tag{21}$$

In order to justify the terms "next greater," "next smaller," it must be shown that an integral u of a homogeneous linear differential equation of the second order

$$\frac{d^2u}{dx^2} + p\frac{du}{dx} + qu = 0$$

can have only a *finite number of zeros* in an interval $(a\,b)$ in which p and q are continuous unless $u \equiv 0$ in $(a\,b)$.

Proof: According to the existence theorem (compare footnote 1, p. 50), u is of class C'' in $(a\,b)$. Suppose u had an infinitude of zeros in $(a\,b)$; then there would exist in $(a\,b)$ at least one accumulation point (compare footnote 1, p. 178) for these zeros. Now either $u(c) \neq 0$; then a vicinity of c can be assigned in which $u(x) \neq 0$. Or else $u(c) = 0$; then $u'(c) \neq 0$ (compare footnote 2, below) and

$$u(c + h) = h\big(u'(c) + (h)\big) \ ;$$

hence a vicinity of c can be assigned in which c is the only zero of $u(x)$. In both cases we reach therefore a contradiction with the assumption that c is an accumulation-point.

The same lemma has to be used, page 108, page 135, page 200, and page 221.

Then it follows from a well-known theorem on homogeneous linear differential equations of the second order due to STURM[2] that every integral of (9) independent of $\Delta(x, x_0)$ vanishes at one and but one point between x_0 and x_0'.

[1] Compare HESSE, *loc. cit.*, p. 448, and A. MAYER, *Journal für Mathematik*, Vol. LXIX (1868), p. 250.

[2] "If u_1, u_2 are two linearly independent integrals of

$$\frac{d^2u}{dx^2} + p\frac{du}{dx} + qu = 0 \ ,$$

We have now to distinguish two cases:

Case I: $x_0' \leqq x_1$.

Then every integral of (9) vanishes at some point of (x_0x_1) and we obtain according to §12, *a*) the

Theorem: If $x_0' \leqq x_1$, *it is possible to make* $\delta^2 J = 0$ *by a proper choice of the function* η.

For instance, by taking $\eta = \Delta(x, x_0)$ in (x_0x_0') and identically zero in $(x_0'x_1)$.

Hence JACOBI inferred that an extremum is impossible if $x_0' \leqq x_1$; for, δJ and $\delta^2 J$ being zero, the sign of ΔJ depends upon the sign of $\delta^3 J$ which can be made negative as well as positive by choosing the sign of ϵ properly. This conclusion is, however, legitimate only after it has been ascertained[1] that the particular variation which causes $\delta^2 J$ to vanish does not at the same time make $\delta^3 J = 0$.

Case II: $x_0' > x_1$ *or else* x_0' *non-existent.*

In this case the particular integral

$$\Delta(x, x_1) = r_1(x) r_2(x_1) - r_2(x) r_1(x_1)$$

of (9) is linearly independent of $\Delta(x, x_0)$ since $\Delta(x_0, x_0) = 0$,

where p and q are functions of x, then between two consecutive zeros of u_1 there is contained one and but one zero of u_2, provided that these zeros are comprised in an interval in which p and q are continuous." See STURM, "Mémoire sur les équations différentielles du second ordre" (*Journal de Liouville*, Vol. I (1836), p. 131); also STURM, *Cours d'Analyse*, 12th ed., Vol. II, No. 609. The theorem follows easily from the well-known formula

$$u_1 \frac{du_2}{dx} - u_2 \frac{du_1}{dx} = Ce^{-\int p\,dx}, \tag{22}$$

where C is a constant $\neq 0$. From the same formula it follows that u_1 and u_2 cannot vanish at the same point, and that u_1 and $\frac{du_1}{dx}$ cannot vanish at the same point.

Compare also DARBOUX, *Théorie des Surfaces*, Vol. III, No. 628, and BÔCHER, *Transactions of the American Mathematical Society*, Vol. II (1901), pp. 150, 428.

It seems that WEIERSTRASS was the first who used Sturm's theorem in this connection. HESSE (*loc. cit.*, p. 257) reaches the same results in a less elegant way by making use of the relation (22).

[1] The value of $\delta^3 J$ for this particular function η has been computed by ERDMANN (*Zeitschrift für Mathematik und Physik*, Vol. XXII (1877), p. 327). He finds, in the notation of §15

$$\delta^3 J = -\epsilon^3 R(x_0')\phi'_\gamma(x_0', \gamma_0)\phi_{\gamma\gamma}(x_0', \gamma_0); \tag{23}$$

$R(x_0')$ and $\phi'_\gamma(x_0', \gamma_0)$ are always different from zero; and $\phi_{\gamma\gamma}(x_0', \gamma_0)$ is also different from zero except when the envelope of the set (28) has a cusp at A' or degenerates into a point. With the exception of these two cases then, JACOBI's result is correct. Compare also §16.

whereas $\Delta(x_0, x_1) = -\Delta(x_1, x_0) \neq 0$.
Hence it follows from Sturm's theorem that $\Delta(x, x_1) \neq 0$ for $x_0 \leqq x < x_1$, and therefore also (on account of the continuity of $\Delta(x, x_1)$) for $x_0 - \delta \leqq x < x_1$, δ being a sufficiently small positive quantity. Now choose x^0 between $x_0 - \delta$ and x_0 and so near to x_0 that[1] $X_0 < x^0 < x_0$. Then we can apply Sturm's theorem to the two particular integrals $\Delta(x, x_1)$ and $\Delta(x, x^0) = r_1(x)\, r_2(x^0) - r_2(x)\, r_1(x^0)$ and obtain the result that

$$\Delta(x, x^0) \neq 0 \quad \text{in } (x_0 x_1) .$$

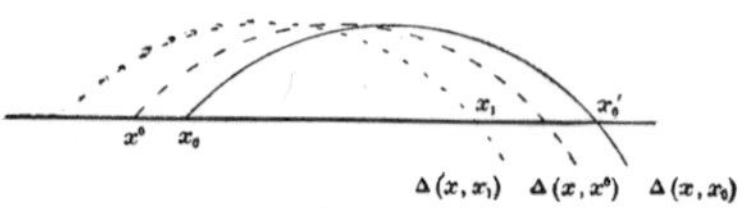

FIG. 9

Simpler proof:

Choose x_2 so that $x_1 < x_2 < x_0'$ and at the same time $x_2 < X_1$ (the quantity introduced on p. 55). Then $\Delta(x, x_2)$ and $\Delta(x, x_0)$ are two linearly independent integrals of (9). Applying STURM's theorem to these two functions we obtain the result that

$$\Delta(x, x_2) \neq 0 \qquad \text{in } (x_0, x_1) .$$

We obtain, therefore, according to §11 *c*), the

Theorem: If $R > 0$ throughout $(x_0 x_1)$, and either $x_1 < x_0'$ or x_0' non-existent, then $\delta^2 J$ is positive for all admissible functions η.

Hence JACOBI inferred that in this case a minimum actually exists, and this was generally believed until WEIERSTRASS showed the fallacy of the conclusion (1879) (see §17).

The above two theorems constitute "JACOBI'S CRITERION." The value x_0' is called *the conjugate of the value x_0;* and the point A' of the extremal $\mathfrak{E}_0$ whose abscissa is x_0', *the conjugate of the point A* whose abscissa is x_0.

§15. GEOMETRICAL INTERPRETATION OF THE CONJUGATE POINTS

JACOBI[2] has given a very elegant geometrical interpretation of the conjugate points, which is based upon the consideration of *the set of extremals through the point A.*

[1] See §13, *a*). On account of (16), $R(x) > 0$ and, therefore, $r_1(x)$ and $r_2(x)$ are continuous not only in $(x_0 x_1)$ but also in the larger interval $(X_0 X_1)$.

[2] *Loc. cit.*, and *Vorlesungen über Dynamik*, p. 46; also HESSE, *loc. cit.*, p. 448.

a) This set is defined by the two equations

$$y = f(x, \alpha, \beta) \ , \tag{24}$$

$$y_0 = f(x_0, \alpha, \beta) \ . \tag{25}$$

The second equation is satisfied by $\alpha = \alpha_0$, $\beta = \beta_0$; and at least one of the two partial derivatives

$$f_\alpha(x_0, \alpha_0, \beta_0) = r_1(x_0) \qquad \text{and} \qquad f_\beta(x_0, \alpha_0, \beta_0) = r_2(x_0)$$

is $\neq 0$ since $r_1(x)$ and $r_2(x)$ are two independent integrals of (9) and $R(x_0) \neq 0$ (see p. 58, footnote 2). According to the theorem[1] on implicit functions we can therefore solve (25) either with respect to α or with respect to β. But we obtain a more symmetrical result if we express α and β in terms of a third parameter γ.

If we choose, for instance,

$$\gamma = f_x(x_0, \alpha, \beta) \tag{26}$$

and denote by γ_0 the value

$$\gamma_0 = f_x(x_0, \alpha_0, \beta_0) \ ,$$

we can solve[2] the two equations (25) and (26) with respect to α and β, and obtain a unique solution

$$\alpha = \alpha(\gamma) \ , \qquad \beta = \beta(\gamma) \ ,$$

which is continuous in the vicinity of the point $\gamma = \gamma_0$ and satisfies the condition

$$\alpha_0 = \alpha(\gamma_0) \ , \qquad \beta_0 = \beta(\gamma_0) \ .$$

Moreover the functions $\alpha(\gamma)$, $\beta(\gamma)$ admit, in the vicinity of of γ_0, continuous first derivatives.

Hence it follows that if we put

$$f(x, \alpha(\gamma), \beta(\gamma)) = \phi(x, \gamma) \ ,$$

the function $\phi(x, \gamma)$, its first partial derivatives and the derivatives[3] ϕ_{xx}, $\phi_{x\gamma}$ will be continuous in the domain

[1] Compare p. 35, footnote 2.

[2] All the conditions of the theorem on implicit functions are fulfilled at the point $\alpha = \alpha_0$, $\beta = \beta_0$, $\gamma = \gamma_0$. In particular, the Jacobian of the two functions $f(x_0, \alpha, \beta) - y_0$ and $f_x(x_0, \alpha, \beta) - \gamma$ with respect to α and β is $\neq 0$ for $\alpha = \alpha_0$, $\beta = \beta_0$, $\gamma = \gamma_0$, its value being $D(x_0) = r_1(x_0)\, r_2'(x_0) - r_2(x_0)\, r_1'(x_0)$, which is different from zero, since r_1, r_2 are linearly independent and x_0 is a non-singular point of the differential equation (9).

[3] Also $\phi_{\gamma\gamma}$ will be continuous if $f_{\alpha\alpha}$, $f_{\alpha\beta}$, $f_{\beta\alpha}$ are continuous in $\mathfrak{A}$.

$$X_0 \leqq x \leqq X_1 \ , \qquad |\gamma - \gamma_0| \leqq d_1 \ ,$$

d_1 being a sufficiently small positive quantity. Furthermore, the equation

$$y_0 = \phi(x_0, \gamma) \tag{27}$$

is satisfied for all sufficiently small values of $|\gamma - \gamma_0|$.

The equation

$$y = \phi(x, \gamma) \tag{28}$$

represents, therefore, the set of extremals through A in a certain vicinity of the extremal $\mathfrak{E}_0$, the latter itself being represented by

$$\mathfrak{E}_0 : \qquad y = \phi(x, \gamma_0) \ . \tag{29}$$

By differentiation with respect to γ we get

$$\phi_\gamma(x, \gamma) = f_\alpha(x, \alpha, \beta)\frac{d\alpha}{d\gamma} + f_\beta(x, \alpha, \beta)\frac{d\beta}{d\gamma} \ ,$$

and therefore, on putting $\gamma = \gamma_0$,

$$\phi_\gamma(x, \gamma_0) = \frac{-r_2(x_0)r_1(x) + r_1(x_0)r_2(x)}{r_1(x_0)r_2'(x_0) - r_2(x_0)r_1'(x_0)} \ .$$

The functions $\phi_\gamma(x, \gamma_0)$ and $\Delta(x, x_0)$ differ, therefore, only by a constant factor:[1]

$$\phi_\gamma(x, \gamma_0) = C\Delta(x, x_0) \ , \qquad C \neq 0 \tag{30}$$

and consequently the *conjugate value* x_0' *may also be defined*[2] *as the root next greater than* x_0 *of the equation*

$$\phi_\gamma(x, \gamma_0) = 0 \ . \tag{30a}$$

From (30) and the properties[3] of $\Delta(x, x_0)$ it follows further that

$$\phi_{\gamma x}(x_0, \gamma_0) \neq 0 \qquad \phi_{\gamma x}(x_0', \gamma_0) \neq 0 \tag{31}$$

Simpler proof of (30): $\phi_\gamma(x, \gamma_0)$ and $\Delta(x, x_0)$ are integrals of JACOBI's differential equation; both vanish for $x = x_0$ without

[1] The same results concerning $\phi(x, \gamma)$ hold if, instead of the particular parameter γ chosen above, we introduce another parameter γ' connected with γ by a relation of the form

$$\gamma = \chi(\gamma') \ ,$$

where $\chi(\gamma)$ and its first derivative are continuous in the vicinity of γ_0, and $\chi'(\gamma_0) \neq 0$.

[2] Compare ERDMANN, *Zeitschrift für Mathematik und Physik*, Vol. XXII (1877), p. 325.

[3] Compare p. 58, footnote 2 and p. 137, footnote 1.

being identically zero. Hence they can differ only by a constant factor.

b) According to the preceding results, the co-ordinates x_0', y_0' of the conjugate point A' satisfy the two equations

$$\Phi\,(x_0', y_0', \gamma_0) \equiv \phi\,(x_0', \gamma_0) - y_0' = 0\ ,$$
$$\Phi_\gamma(x_0', y_0', \gamma_0) \equiv \phi_\gamma(x_0', \gamma_0) \qquad = 0\ ,$$

and the determinant

$$\begin{vmatrix} \Phi_x & \Phi_y \\ \Phi_{\gamma x} & \Phi_{\gamma y} \end{vmatrix}$$

is different from zero for $x = x_0'$, $y = y_0'$, $\gamma = \gamma_0$, its value being $\phi_{\gamma x}(x_0', \gamma_0)$. Hence we obtain, according to the theory of envelopes,[1] the following *geometrical interpretation:*

Consider the extremal

$$\mathfrak{E}_0: \qquad y = \phi(x, \gamma_0)$$

and a neighboring extremal of the set (28):

$$\mathfrak{E}: \qquad y = \phi\,(x, \gamma_0 + k)\ .$$

Then if $|k|$ be chosen sufficiently small, the curve $\mathfrak{E}$ will meet $\mathfrak{E}_0$ at one and but one point P in the vicinity[2] of A'. And as k approaches zero, the point P approaches A' as limiting position. Hence we have the

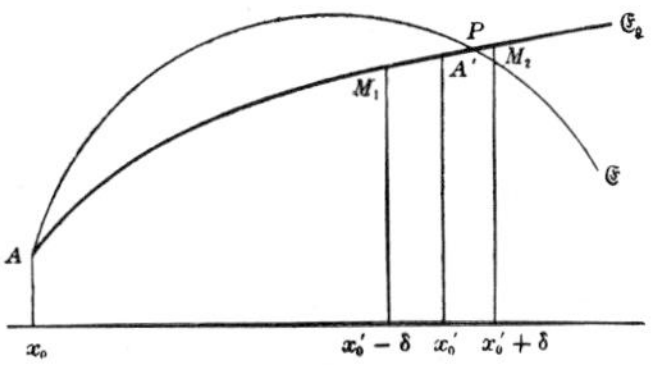

FIG. 10

Theorem: The conjugate A' of the point A is the point where the extremal $\mathfrak{E}_0$ touches for the first time the envelope of the set of extremals through A.

c) EXAMPLE IV: $F = g(y')$, a function of y' alone.

[1] Compare E. III D, p. 47. The proof presupposes the continuity of $\Phi_x, \Phi_y, \Phi_\gamma, \Phi_{\gamma x}, \Phi_{\gamma y}, \Phi_{\gamma\gamma}$ in the vicinity of the point $x = x'_0$, $y = y'_0$, $\gamma = \gamma_0$. These conditions are satisfied in our case provided that x_0' lies in the interval $(X_0 X_1)$, and provided that we suppose that not only the derivatives mentioned on p. 55, but also $f_{\alpha\alpha}, f_{\alpha\beta}, f_{\beta\beta}$ are continuous in $\mathfrak{A}$ (compare p. 54, footnote 3).

[2] This means: If we choose a positive quantity δ arbitrarily but sufficiently small, and denote by M_1 and M_2 the points of $\mathfrak{E}_0$ whose abscissae are $x_0' - \delta$ and $x_0' + \delta$ then another positive quantity σ can be determined such that every extremal $\mathfrak{E}$ for which $|k| < \sigma$ meets $\mathfrak{E}_0$ at one and but one point P between M_1 and M_2. Compare p. 35, footnote 2.

The extremals are straight lines; the set of extremals (28) is the pencil of straight lines through A; hence there exists no conjugate point.

The same result follows analytically: The general solution of Euler's equation is $y = \alpha x + \beta$, hence

$$r_1 = x \ , \qquad r_2 = 1 \ ,$$

and

$$\Delta(x, x_0) = x - x_0 \ .$$

EXAMPLE I (see p. 27): From the general solution of Euler's equation

$$y = \alpha \cosh \frac{x - \beta}{\alpha}$$

we get

$$\Delta(x, x_0) = \sinh v \cosh v_0 - \sinh v_0 \cosh v + (v - v_0) \sinh v \sinh v_0 \ ,$$

where

$$v = \frac{x - \beta_0}{\alpha_0} \ , \qquad v_0 = \frac{x_0 - \beta_0}{\alpha_0} \ .$$

Hence we obtain (if $v_0 \neq 0$) for the determination of x_0' the transcendental equation

$$\coth v - v = \coth v_0 - v_0 \ . \tag{32}$$

Since the function $\coth v - v$ decreases from $+\infty$ to $-\infty$ as v increases from $-\infty$ to 0, and from $+\infty$ to $-\infty$ as v increases from 0 to $+\infty$, the equation (32) has, besides the trivial solution $v = v_0$, one other solution v_0', and v_0 and v_0' have opposite signs.

Hence if $v_0 > 0$, *i. e., if A lies on the ascending branch of the catenary, there exists no conjugate point:* $\Delta(x, x_0) \neq 0$ for every $x > x_0$. The same result follows for $v_0 = 0$.

If, on the contrary, $v_0 < 0$, *i. e.*, if A lies on the descending branch of the catenary, there always exists a conjugate point A' situated on the ascending branch. It can be determined geometrically by the following property, discovered by *Lindelöf*:[1] *The tangents to the catenary at A and at A' meet on the x-axis.*

If, on the contrary, x_2 be any value in the interval (X_0X_1) for which

$$\phi_\gamma(x_2, \gamma_0) \neq 0 \ ,$$

then two positive quantities δ' and σ' can be determined such that no extremal $\mathfrak{E}$ for which $|k| < \sigma'$ meets $\mathfrak{E}_0$ between the points whose abscissae are $x_2 - \delta'$ and $x_2 + \delta'$.

For in this case the difference

$$\phi(x_2 + h, \gamma_0 + k) - \phi(x_2 + h, \gamma_0) = k\phi_\gamma(x_2 + h, \gamma_0 + \theta k) \ ,$$

where $0 < \theta < 1$ is different from zero for all sufficiently small values of $|h|$ and $|k|$.

[1] LINDELÖF-MOIGNO, *loc. cit.*, p. 209, and LINDELÖF, *Mathematische Annalen*, Vol. II (1870), p. 160. Compare also the references given on p. 28, footnote 1.

For the abscissae of the points of intersection of these two tangents with the x-axis are

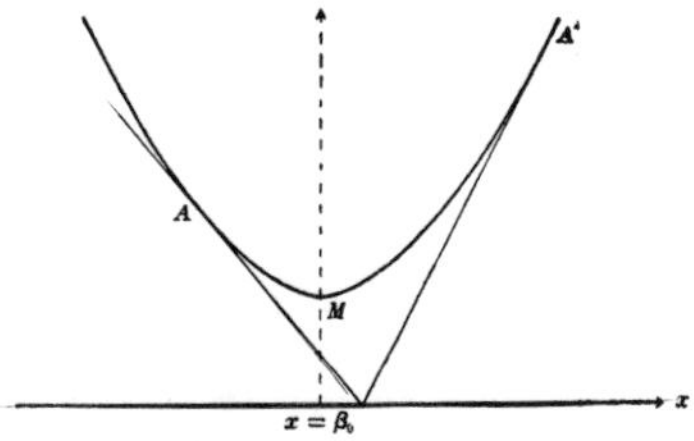

FIG. 11

$$X = x_0 - \alpha_0 \coth \frac{x_0 - \beta_0}{\alpha_0},$$

and

$$X' = x_0' - \alpha_0 \coth \frac{x_0' - \beta_0}{\alpha_0},$$

and they are equal on account of (32).

§16. NECESSITY OF JACOBI'S CONDITION

It has already been pointed out that the two theorems of §14 which constitute Jacobi's Criterion, though giving important information concerning the sign of the second variation, contain neither a necessary nor a sufficient condition for a minimum or maximum.

But at least a *necessary* condition can be derived from the first of the two theorems by a slight modification of the reasoning: If $x_0' < x_1$, then $\delta^2 J$ can be made not only zero but *even negative.*

This was first proved by WEIERSTRASS in his lectures; the first *published* proof is due to ERDMANN.[1] The following is essentially Erdmann's proof:

[1] *Zeitschrift für Mathematik und Physik*, Vol. XXIII (1878), p. 367. SCHEEFFER'S proof (*Mathematische Annalen*, Vol. XXV (1885), p. 548), is not essentially different from ERDMANN'S.

WEIERSTRASS writes the second variation in the form

$$\delta^2 J = \epsilon^2 \left\{ \int_{x_0}^{x_1} [(P+k)\eta^2 + 2Q\eta\eta' + R\eta'^2]\,dx - k\int_{x_0}^{x_1} \eta^2 dx \right\},$$

k being a small positive constant, and applies to the first integral *Jacobi's* transformation:

$$\delta^2 J = \epsilon^2 \left\{ \int_{x_0}^{x_1} \eta\bar{\Psi}(\eta)\,dx - k\int_{x_0}^{x_1} \eta^2 dx \right\},$$

where

$$\bar{\Psi}(\eta) = \left((P+k) - Q'\right)\eta - \frac{d}{dx}(R\eta').$$

Then he shows that there exist admissible functions η which satisfy the differential equation $\bar{\Psi}(\eta) = 0$. For such a function η, $\delta^2 J$ is evidently negative.

H. A. SCHWARZ (*Lectures*, 1898–99) uses the following function η:

$$\eta = \begin{cases} \Delta(x, x_0) + k\omega & \text{in } (x_0 x_0'), \\ k\omega & \text{in } (x_0' x_1), \end{cases}$$

Take x_2' so that

$$x_0' < x_2' < x_1 \quad \text{and} \quad \Delta(x_2', x_0) \neq 0 ,$$

and put

$$u = \Delta(x, x_0) ,$$
$$v = \rho\Delta(x, x_2') ,$$

where $\rho = +1$ or -1; u and v are particular integrals of (9) and linearly independent; hence the relation (22) holds and takes the following form for the differential equation (9):

$$R(uv' - u'v) = K , \tag{33}$$

K being a constant different from zero.

We choose ρ so that $K > 0$; this is always possible, for, if v is replaced by $-v$, K is changed into $-K$.

Further, since also u and $u - v$ are linearly independent, it follows from Sturm's theorem (see p. 58, footnote 2) that $u - v$ vanishes for one value of x, say $x = c$, between x_0 and x_0'; hence

$$u(c) = v(c) .$$

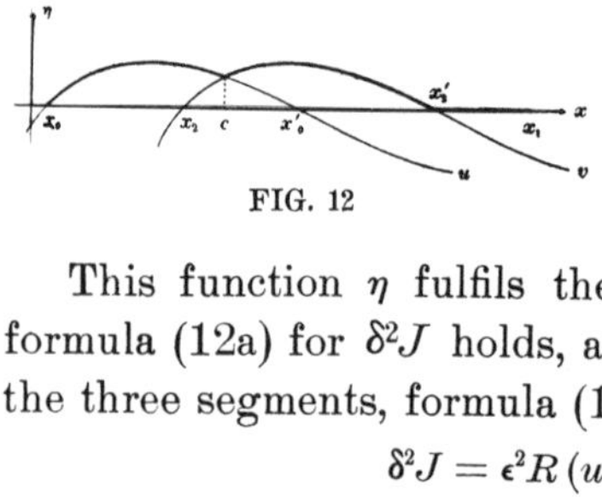

FIG. 12

Now define η as follows:

$$\eta = \begin{cases} u & \text{in } (x_0\, c) , \\ v & \text{in } (c\; x_2') , \\ 0 & \text{in } (x_2' x_1) . \end{cases}$$

This function η fulfils the conditions under which the formula (12a) for $\delta^2 J$ holds, and since $\Psi(\eta) = 0$ for each of the three segments, formula (12a) becomes:

$$\delta^2 J = \epsilon^2 R(uu' - vv')\,\Big|^{c} ,$$

where k is a small constant and ω is a function of class C'' which vanishes at x_0 and x_1 but not at x_0'. The corresponding value of $\delta^2 J$ is of the form:

$$\delta^2 J = \epsilon^2 \left\{ 2k R(x_0') \Delta'(x_0', x_0)\, \omega(x_0') + k^2 V \right\} ,$$

which can be made negative by a proper choice of k. (Compare SOMMERFELD, *Jahresbericht der Deutschen Mathematiker-Vereinigung*, Vol. VIII (1900), p. 189.)

All these proofs presuppose $x_0' < x_1$; for the case $x_0' = x_1$, so far as it is not covered by Erdmann's formula (23) for $\delta^3 J$, compare KNESER, *Mathematische Annalen*, Vol. L (1897), p. 50, and OSGOOD, *Transactions of the American Mathematical Society*, Vol. II (1901), p. 166. This case will be treated in parameter-representation in chap. v, § 38.

which may be written, since $u(c)=v(c)$:

$$\delta^2 J = -\epsilon^2 R(uv' - u'v)\,|^c = -\epsilon^2 K \ ,$$

and this is negative according to our agreements concerning the sign of v.

Thus we have proved the

FUNDAMENTAL THEOREM III: *The third necessary condition for a minimum (maximum) is that*

$$\Delta(x, x_0) \neq 0 \qquad \text{(III)}$$

for all values of x in the open interval $x_0 < x < x_1$.

Corollary: The same condition may also be written

$$x_1 \leqq x_0', \text{ or else } x_0' \text{ non-existent} \ , \qquad \text{(III)}$$

i. e., if the end-point B lies beyond the conjugate point A', there is no minimum or maximum.

We shall refer to this condition as JACOBI'S *condition.*

CHAPTER III

SUFFICIENT CONDITIONS

§17. SUFFICIENT CONDITIONS FOR A "WEAK MINIMUM"[1]

WE suppose henceforth that for our extremal $\mathfrak{E}_0$ the conditions

$$R > 0 \tag{II'}$$

$$\Delta(x, x_0) \neq 0 \quad \text{for} \quad x_0 < x \leqq x_1 \,{}^{2} \tag{III'}$$

are fulfilled, and we ask: *Are these conditions* SUFFICIENT *for a minimum?*

a) It seems so, and until rather recently it was generally believed to be so: For the reasoning of §11 shows that after an admissible function η has been chosen, ΔJ will be positive for all sufficiently small values of $|\epsilon|$; hence within the set of curves with parameter ϵ:

$$\overline{y} = y + \epsilon\eta \tag{1}$$

the curve $\mathfrak{E}_0$ does furnish a minimum. On the other hand, every curve $\overline{\mathfrak{C}}$ may be considered as an individual of such a set, and therefore it seems as if we must actually have a minimum.

But a closer analysis shows that the conclusion is wrong. For all we have proved so far is this: After a function η has been selected we can assign a positive quantity[3] ρ_η such that $\Delta J > 0$ for every $|\epsilon| < \rho_\eta$. And if

[1] Compare for this section SCHEEFFER, "Ueber die Bedeutung der Begriffe Maximum und Minimum in der Variationsrechnung," *Mathematische Annalen*, Vol. XXVI (1886), p. 197. This paper has been of the greatest importance in clearing up the fundamental conceptions in the Calculus of Variations.

[2] Notice the equality sign which distinguishes (III') from (III); for the case $x_1 = x_0'$, which we omit here, compare the references on p. 65, footnote.

[3] The notation ρ_η indicates that ρ depends on the function η; compare E. H. MOORE, *Transactions of the American Mathematical Society*, Vol. I (1900), p. 500.

we denote by m_η the maximum of $|\eta|$ in $(x_0 x_1)$ and put $k_\eta = m_\eta \rho_\eta$, we have

$$|\Delta y| < k_\eta$$

for all curves of the set (1) for which $|\epsilon| < \rho_\eta$; and *vice versa*, if we draw in the neighborhood (k_η) of $\mathfrak{C}_0$ any curve of this particular set, the corresponding ϵ satisfies the inequality $|\epsilon| < \rho_\eta$ and therefore $\Delta J > 0$.

Now consider the totality of all admissible functions η; the corresponding set of values k_η has a lower limit $k_0 \geqq 0$. If it could be proved that $k_0 > 0$, then we could infer that $\Delta J > 0$ for every admissible variation $\bar{y}$ for which $|\Delta y| < k_0$, and we would actually have a minimum. But it cannot be proved that $k_0 > 0$ and therefore we cannot infer that $\mathfrak{C}_0$ minimizes J.

It is even *a priori* clear that the method which we have followed so far can never lead to a proof of the sufficiency of this or any other set of conditions.[1]

For, if we apply Taylor's expansion (either infinite or with the remainder term) to the difference

$$\Delta F = F(x, y + \Delta y, y' + \Delta y') - F(x, y, y')$$

and integrate, we can only draw conclusions concernig the sign of ΔJ from the sign of the first terms, if *not only* $|\Delta y|$ *but also* $|\Delta y'|$ *remains sufficiently small*, or geometrically: if for corresponding points of $\mathfrak{C}_0$ and $\bar{\mathfrak{C}}$ not only the distance but also the difference of the directions of the tangents is sufficiently small.

b) If there exists a positive quantity k such that $\Delta J \geqq 0$ for all admissible variations for which

$$|\Delta y| < k \qquad \textit{and} \qquad |\Delta y'| < k \; ,$$

KNESER (*Lehrbuch*, §17) says that the curve $\mathfrak{C}_0$ furnishes a "*Weak Minimum*," from which he distinguishes the mini-

[1] First emphasized by WEIERSTRASS.

mum as we have defined[1] it according to WEIERSTRASS, as "*Strong Minimum*." If a curve furnishes a strong minimum, it always furnishes *a fortiori* also a weak minimum, but not *vice versa*.

If we adopt temporarily this terminology, we can enunciate the following

Theorem: An extremal $\mathfrak{E}_0$ *for which the conditions*

$$R > 0 \tag{II'}$$

$$\Delta(x, x_0) \neq 0 \qquad \text{for} \qquad x_0 < x \leqq x_1 \tag{III'}$$

are fulfilled, furnishes at least a "weak minimum" for the integral J.

The first proof of this theorem was given by WEIERSTRASS (*Lectures*, 1879), the first published proof by SCHEEFFER (*loc. cit.*, 1886). The following proof is due to KNESER:[2]

We return to equation (3) of §11 which we write in the form:

$$\Delta J = \tfrac{1}{2}\int_{x_0}^{x_1} (P\omega^2 + 2\,Q\omega\omega' + R\omega'^2)\,dx + \tfrac{1}{2}\int_{x_0}^{x_1} (L\omega^2 + N\omega'^2)\,dx \ ,$$

where $\omega = \Delta y$, and L, N are infinitesimals in the following sense: corresponding to every positive quantity σ another positive quantity ρ_σ can be assigned such that:

$$|L| < \sigma \ , \qquad |N| < \sigma \qquad \text{in} \qquad (x_0 x_1) \ ,$$

provided that

$$|\omega| < \rho_\sigma \qquad \text{and} \qquad |\omega'| < \rho_\sigma \qquad \text{in} \qquad (x_0 x_1) \ .$$

By Legendre's transformation,[3] the first integral may be thrown into the form:

$$\tfrac{1}{2}\int_{x_0}^{x_1}\left[R\left(\omega' + \frac{Q+w}{R}\,\omega\right)^2 + \left(P + w' - \frac{(Q+w)^2}{R}\right)\omega^2\right]dx \ .$$

1 Compare §3, *b*).

2 *Jahresbericht der Deutschen Mathematiker-Vereinigung*, Vol. VI (1899), p. 95. The theorem can also be proved by means of Weierstrass's Theorem (§20); compare KNESER, *Lehrbuch*, §§20–22.

3 Compare §11, *b*).

Since the conditions (II′) and (III′) are fulfilled, there exist[1] solutions of the differential equation

$$P + \frac{dw}{dx} - \frac{(Q+w)^2}{R} = 0 \ ,$$

which are of class C' in (x_0x_1); hence it follows[2] that, provided the constant c be taken sufficiently small, there also exist integrals of the differential equation

$$P + \frac{dw}{dx} - \frac{(Q+w)^2}{R} = c^2 \ , \tag{2}$$

which are of class C' in (x_0x_1); let w be such an integral, and introduce

$$\zeta = \omega' + \frac{Q+w}{R}\omega$$

instead of ω'. Then ΔJ takes the form

$$\Delta J = \tfrac{1}{2}\int_{x_0}^{x_1} \left[(c^2+\lambda)\,\omega^2 + 2\mu\omega\zeta + (R+\nu)\,\zeta^2 \right] dx \ ,$$

where λ, μ, ν are infinitesimals in the same sense as L and N. But this may be written

$$\Delta J = \tfrac{1}{2}\int_{x_0}^{x_1} \left[(R+\nu)\left(\zeta + \frac{\mu}{R+\nu}\omega\right)^2 + \left(c^2 + \lambda - \frac{\mu^2}{R+\nu}\right)\omega^2 \right] dx \ ,$$

and since λ, μ, ν are infinitesimals, we can choose a positive quantity k so that $R+\nu > 0$ and $c^2 + \lambda - \mu^2/(R+\nu) > 0$ in (x_0x_1), and consequently $\Delta J > 0$, provided that $|\omega| < k$ and $|\omega'| < k$, Q. E. D.

Remark: We have given this theorem chiefly for its historical interest: It marks the farthest point which the Calculus of Variations had reached before WEIERSTRASS'S

[1] This follows from the connection between Legendre's and Jacobi's differential equations; see equation (8) in §11, *b*).

[2] According to a theorem due to POINCARÉ (*Mécanique celeste*, Vol. I, p. 58; compare also E. II A, p. 205, and PICARD, *Traité*, etc., Vol. III, p. 169). A similar theorem was given by WEIERSTRASS in his lectures in connection with his proof of the necessity of Jacobi's condition, see p. 65, footnote.

epoch-making discoveries concerning the sufficient conditions for a "strong minimum."

After these discoveries, only a secondary importance attaches itself to the "weak minimum;" for the restriction imposed upon the derivative in the "weak minimum" is indeed a very artificial[1] one, only suggested and justified by the former inability of the Calculus of Variations to dispense with it.

c) The terms "weak" and "strong" are sometimes also applied to the variations. *A variation containing a parameter* ϵ

$$\Delta y = \omega(x, \epsilon)$$

is called *weak* if not only

$$\lim_{\epsilon=0} \omega(x, \epsilon) = 0 \quad \text{but also} \quad \lim_{\epsilon=0} \omega_x(x, \epsilon) = 0$$

uniformly in (x_0x_1), *strong* if this condition is not satisfied.

The variations of the form

$$\Delta y = \epsilon\eta ,$$

as well as the more general variations which we have mentioned in §4, *d*), are weak variations.

WEIERSTRASS gives the following example[2] of a strong variation:

$$\Delta y = \epsilon \sin\left(\frac{(x - x_0)\pi}{\epsilon^n}\right) ,$$

n a positive integer; here the condition

$$\lim_{\epsilon=0} \Delta y = 0$$

[1] Especially if we think of geometrical problems, for instance, the problem of the shortest curve on a given surface between two points.

For the more general problem, however, where higher derivatives occur under the integral sign, such restrictions are of greater importance; compare ZERMELO, *Dissertation*, pp. 26–31.

[2] The following modification of WEIERSTRASS's example has the advantage of vanishing at both end-points:

$$\Delta y = \frac{1}{m} \sin\left(\frac{(x - x_0)m^n\pi}{x_1 - x_0}\right) ,$$

m and n being positive integers.

is satisfied, but not the condition

$$\underset{\epsilon=0}{L}\,\Delta y' = 0 \ .$$

Other examples of strong variations will occur in §§18 and 22.

§18. INSUFFICIENCY OF THE PRECEDING CONDITIONS FOR A STRONG MINIMUM, AND FOURTH NECESSARY CONDITION

From the introductory remarks of the previous section, it follows that we have no reason to expect that the conditions (I), (II′), (III′) are sufficient for a minimum in the sense in which we have defined it according to WEIERSTRASS (a "strong minimum" in KNESER'S terminology).

a) As a matter of fact *the three conditions* (*I*), (*II′*) and (*III′*) *are* NOT *sufficient for a strong minimum*, and it is easy to construct examples[1] which prove this statement:

EXAMPLE III[2] (see p. 39):

$$F = y'^2 (y' + 1)^2 \ .$$

Here $\mathfrak{E}_0$ is the straight line joining the two given points A and B, say

$$\mathfrak{E}_0: \qquad y = mx + n \ .$$

Further:

$$R = 2\,(6m^2 + 6m + 1) \ ,$$

$$\Delta\,(x, x_0) = x - x_0 \ ;$$

hence x_0' non-existent. Let m_1, m_2 be the two roots of the equation

$$6m^2 + 6m + 1 = 0 \ , \text{ viz.,}$$

$$m_1 = \tfrac{1}{2}\left(-1 + \frac{1}{\sqrt{3}}\right) = -0\cdot 2113\cdots$$

[1] The first example of this kind was the problem of the solid of revolution of least resistance; already LEGENDRE had shown that the resistance can be made as small as we please by a properly chosen zigzag line; see LEGENDRE, *loc. cit.*, p. 73, in STÄCKEL'S translation, and PASCAL, *loc. cit.*, p. 113.

[2] Compare BOLZA, "Some Instructive Examples in the Calculus of Variations," *Bulletin of the American Mathematical Society* (2), Vol. IX (1902), p. 3.

$$m_2 = \tfrac{1}{2}\left(-1 - \frac{1}{\sqrt{3}}\right) = -\,0\cdot7887\cdots$$

then
$$R > 0 \quad \text{if} \quad m > m_1 \quad \text{or} \quad m < m_2\ ,$$
$$R < 0 \quad \text{if} \quad m_2 < m < m_1\ .$$

In the former case, the first three necessary conditions for a minimum, in the latter for a maximum, are satisfied. Nevertheless, if

$$-1 < m < 0\ ,$$

neither a maximum nor a minimum takes place. For, in this case, if any neighborhood (ρ) of $\mathfrak{E}_0$ be given, however small, we can always join A and B by a broken line $\overline{\mathfrak{C}}$ made up of segments of straight lines of slope 0 and -1, and contained in (ρ). But for such a broken line $\bar{J}=0$, whereas for $\mathfrak{E}_0$ the integral J is positive. This proves that $\mathfrak{E}_0$ cannot furnish a minimum. That it cannot furnish a maximum will be seen later, in §18, e).

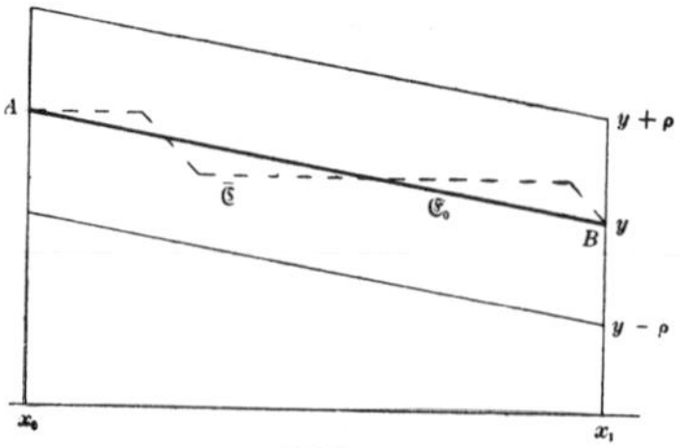

FIG. 13

EXAMPLE V: To minimize

$$J = \int_{x_0}^{x_1} (y'^2 + y'^3)\, dx\ ,$$

the given end-points having the co-ordinates $(x_0, y_0) = (0, 0)$, $(x_1, y_1) = (1, 0)$.

The extremals are straight lines, and $\mathfrak{E}_0$ is the segment (0 1) of the x-axis. Further,

$$R = 2\ ,$$
$$\Delta(x, x_0) = x - x_0\ .$$

Hence the conditions (I), (II′), III′) for a minimum are satisfied. Nevertheless ΔJ can be made negative. For, if we choose for $\overline{\mathfrak{C}}$ the broken line APB, the co-ordinates of P being $(1-p, q)$, where $0 < p < 1$, and $q > 0$, we obtain

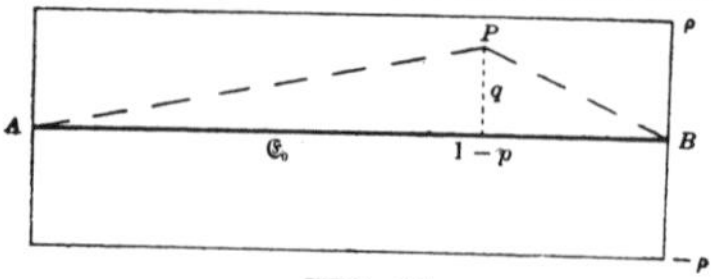

FIG. 14

$$\Delta J = \frac{q^2}{p(1-p)}\left(1 + \frac{q}{1-p} - \frac{q}{p}\right).$$

Any neighborhood (ρ) of $\mathfrak{C}_0$ being given, choose $q < \rho$; then p can always be taken so small that $\Delta J < 0$.

b) The insufficiency of the preceding three conditions being thus established, further conditions must be added before we can be certain that the curve $\mathfrak{C}_0$ minimizes the integral J.

A *fourth necessary condition* was discovered by WEIERSTRASS in 1879 and derived by him in the following manner:

Through an arbitrary point $2 : (x_2, y_2)$ of $\mathfrak{C}_0$ we draw arbitrarily a curve $\tilde{\mathfrak{C}} : \tilde{y} = \tilde{f}(x)$, of class C'.

Denoting by 4 that point of $\tilde{\mathfrak{C}}$ whose abscissa is $x_2 - h$, h being a small positive quantity, we draw, as in §8, a curve $\bar{\mathfrak{C}} : \bar{y} = y + \epsilon\eta$ of class C' from 0 to 4 and replace the arc 02 of $\mathfrak{C}_0$ by the curve 042.

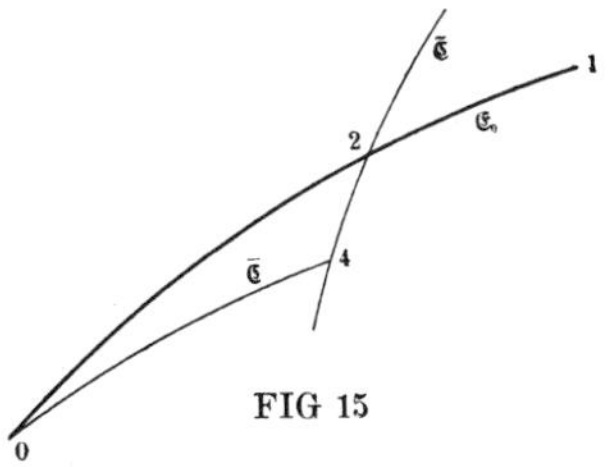

FIG 15

By taking h sufficiently small we can make the curve 042 lie in the neighborhood (ρ) of $\mathfrak{C}_0$.

For this variation of $\mathfrak{C}_0$ we obtain in the notation of §8:

$$\Delta J = \bar{J}_{04} + \tilde{J}_{42} - J_{02} \; . \tag{3}$$

But according to §8, equation (30), this is equal to

$$\Delta J = h\,\mathbf{E}(x_2, y_2;\; y_2', \tilde{y}_2') + h(h) \; , \tag{4}$$

where (h) denotes as usual an infinitesimal, and the **E**-function is defined by

$$\mathbf{E}(x, y;\, p, \tilde{p}) = F(x, y, \tilde{p}) - F(x, y, p) - (\tilde{p} - p) F_{y'}(x, y, p) \, .$$

Hence follows the

FUNDAMENTAL THEOREM IV: *The fourth necessary condition for a minimum (maximum) is that*

$$\mathbf{E}(x, y; y', \tilde{p}) \geqq 0 \ (\leqq 0) \tag{IV}$$

along[1] *the curve* $\mathfrak{E}_0$ *for every finite value of* $\tilde{p}$.

We shall refer to this condition as WEIERSTRASS's *condition.*

c) Applying Taylor's formula to the difference

$$F(x, y, \tilde{p}) - F(x, y, p) ,$$

we obtain the following important *relation*[2] *between the* **E**-*function and* $F_{y'y'}$:

$$\mathbf{E}(x, y; p, \tilde{p}) = \frac{(\tilde{p}-p)^2}{2} F_{y'y'}(x, y, p^*) \tag{5}$$

where

$$p^* = p + \theta(\tilde{p} - p) , \qquad 0 < \theta < 1 .$$

This proves

Corollary I: Condition (IV) *is always satisfied if for every point* (x, y) *on* $\mathfrak{E}_0$ *and for every finite value of* $\tilde{p}$

$$F_{y'y'}(x, y, \tilde{p}) \geqq 0 . \tag{IIa}$$

Furthermore, if we define the function[3] $\mathbf{E}_1(x, y; p, \tilde{p})$ by the equation

$$\mathbf{E}_1(x, y; p, \tilde{p}) = \frac{\mathbf{E}(x, y; p, \tilde{p})}{(\tilde{p}-p)^2} \tag{6}$$

when $\tilde{p} \neq p$, and by

$$\mathbf{E}_1(x, y; p, p) = \underset{\tilde{p}=p}{L}\, \mathbf{E}_1(x, y; p, \tilde{p}) = \tfrac{1}{2} F_{y'y'}(x, y, p) \tag{6a}$$

when $\tilde{p} = p$, we obtain

Corollary II: Condition (IV) *is equivalent to the condition*

$$\mathbf{E}_1(x, y; y', \tilde{p}) \geqq 0 \tag{IVa}$$

along $\mathfrak{E}_0$ *for every finite* $\tilde{p}$.

d) ZERMELO[4] has given the following geometrical

[1] *I. e.*, if (x, y) is any point of $\mathfrak{E}_0$ and y' the slope of $\mathfrak{E}_0$ at (x, y).

[2] Due to ZERMELO, *loc. cit.*, p. 67.

[3] Compare ZERMELO, *loc. cit.*, p. 60.

[4] *Loc. cit.*, p. 67.

interpretation of the relation between the **E**-function and $F_{y'y'}$:

Let $F(p)$ denote the function $F(x, y, p)$ considered as a function of p alone, x, y being regarded as constant, and consider the curve

$$u = F(p) . \tag{7}$$

Draw the tangent P_0T at the point P_0 whose abscissa is $p = y'$; and let P and Q be the points of intersection with the line $p = \tilde{p}$ of the curve and of the tangent P_0T respectively.

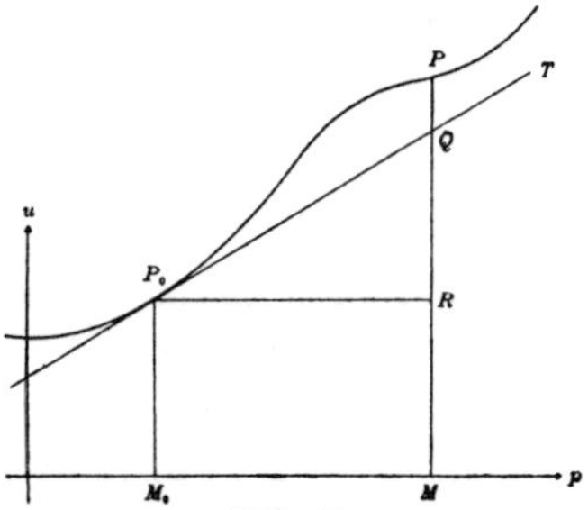

FIG. 16

Then

$$\mathbf{E}(x, y;\ y', \tilde{p}) = F(\tilde{p}) - F(y') - (\tilde{p} - y')\,F'(y')$$

is represented by the vector QP, and the condition

$$\mathbf{E}(x, y;\ y', \tilde{p}) \geqq 0 \tag{IV}$$

means therefore geometrically that *the curve* (7) *lies entirely above—or at least not below—the tangent* P_0T.

In order that (IV) may hold it is therefore:

a) Necessary that the curve shall turn its convex side downward *at* $p = y'$, *i. e.*, that

$$F''(y') \geqq 0 .$$

This is our old condition (II), which is consequently contained in the new condition (IV).

β) Sufficient that the curve shall everywhere turn its convex side downward, *i. e.*, that

$$F''(p) \geqq 0$$

for every p, which is the above condition (IIa).

But neither is the first condition sufficient, nor the second necessary.

e) EXAMPLE I (see p. 49):

$$F = y\sqrt{1 + y'^2} ;$$

hence

$$F_{y'y'}(x, y, p) = \frac{y}{\left(\sqrt{1+p^2}\right)^3} \; .$$

Since $y > 0$ along the catenary, condition (IIa), and therefore also (IV), is satisfied.

EXAMPLE III (see pp. 39, 73):

$$F = y'^2 (y' + 1)^2 \; ;$$

hence

$$\mathbf{E}(x, y; y', \tilde{p}) = (\tilde{p} - y')^2 \left[\tilde{p}^2 + 2\tilde{p}(y'+1) + 3y'^2 + 4y' + 1\right] .$$

$\mathfrak{C}_0$ is the straight line joining the two points 0 and 1, say: $y = mx + n$; hence along $\mathfrak{C}_0$, $y' = m$.

The quadratic in $\tilde{p}$

$$\tilde{p}^2 + 2\tilde{p}(m+1) + 3m^2 + 4m + 1$$

is always positive if $m(m+1) > 0$; it can change sign if $m(m+1) < 0$; and it reduces to a complete square if $m(m+1) = 0$.

Hence we obtain the result:

If $m \geqq 0$ or $m \leqq -1$, condition (IV) is satisfied; if $-1 < m < 0$, condition (IV) is not satisfied, and the line 01 furnishes no extremum, in accordance with the results of §18, a).

EXAMPLE V (see p. 74):

$$F = y'^2 + y'^3 \; ;$$

hence along the curve $\mathfrak{C}_0 : y = 0$ we have

$$\mathbf{E}(x, y; y', \tilde{p}) = \tilde{p}^2 (1 + \tilde{p}) \; ,$$

which can change sign at every point of $\mathfrak{C}_0$. Condition (IV) is therefore not satisfied.

§19. EXISTENCE OF A "FIELD OF EXTREMALS"

Before we can take up the question of sufficient conditions, we must introduce the important concept of a "field of extremals."

a) Definition of a "field."

Consider any one-parameter set of extremals[1]

$$y = \phi(x, \gamma) \; , \tag{8}$$

[1] Here the symbol $\phi(x, \gamma)$ is used in a more general sense than in §15.

in which our extremal $\mathfrak{E}_0$ is contained, say for $\gamma=\gamma_0$. Suppose $\phi(x, \gamma)$, its first partial derivatives and the derivatives $\phi_{xx}, \phi_{x\gamma}$ to be continuous functions of x and γ in the domain

$$X_0 \leqq x \leqq X_1 , \qquad |\gamma - \gamma_0| \leqq d_0 ,$$

d_0 being a positive quantity and X_0, X_1 having the same signification as in §13. Let k denote a positive quantity less than d_0, and $\mathfrak{S}_k$ the set of points (x, y) furnished by (8) as x and γ take all the values in the domain

$$\mathfrak{B}_k : \qquad x_0 \leqq x \leqq x_1 , \qquad |\gamma - \gamma_0| \leqq k .$$

$\mathfrak{S}_k$ may also be defined as the strip of the x, y-plane swept out by the extremals (8) as γ increases from γ_0-k to γ_0+k, x being restricted to the interval (x_0x_1).

Then $\mathfrak{S}_k$ is called[1] a "*field of extremals about the arc* $\mathfrak{E}_0$" *if through every point* (x, y) *of* $\mathfrak{S}_k$ *there passes* BUT ONE EXTREMAL *of the set* (8) *for which* $|\gamma-\gamma_0|\leqq k$.

This means analytically that there exists a single-valued function

$$\left.\begin{aligned} \gamma &= \psi(x, y) \\ y &= \phi\big(x, \psi(x, y)\big) \end{aligned}\right\} , \tag{9}$$

such that

and

$$|\psi(x, y) - \gamma_0| \leqq k$$

for every (x, y) in $\mathfrak{S}_k$.

In addition to this principal property we shall include in the definition of a field the further conditions that the inverse function $\psi(x, y)$ shall be of class C' in $\mathfrak{S}_k$, and that it shall be possible to choose a positive quantity ρ so small that the domain $\mathfrak{S}_k$ contains the neighborhood (ρ) of the extremal $\mathfrak{E}_0$.

b) With respect to *the existence of a field* the following theorem holds:

Whenever

$$\phi_\gamma(x, \gamma_0) \neq 0 \qquad \textit{throughout} \quad (x_0x_1) , \tag{10}$$

[1] According to KNESER, *Lehrbuch*, §14; the notion of a field is due, in a more special sense, to WEIERSTRASS; in its most general sense to H. A. SCHWARZ, *Werke*, Vol. I, p. 225. Compare also OSGOOD, *loc. cit.*, p. 112.

k can be taken so small that the extremals (8) *furnish a field* $\mathfrak{S}_k$ *about* $\mathfrak{E}_0$.

Proof:[1] From (10) it follows that $\phi_\gamma(x, \gamma_0)$—being continuous in (x_0x_1)—cannot change sign in (x_0x_1). In order to fix the ideas suppose that

$$\phi_\gamma(x, \gamma_0) > 0 \qquad \text{in } (x_0x_1) .$$

Then it follows, according to well-known theorems[2] on continuous functions, that k can be taken so small that

$$\phi_\gamma(x, \gamma) > 0 \qquad \text{in } \mathfrak{B}_k . \tag{11}$$

Hence if we give x any fixed value x_2 contained in (x_0x_1) and let γ increase from $\gamma_0 - k$ to $\gamma_0 + k$, $\phi(x_2, \gamma)$ increases continually from $\phi(x_2, \gamma_0 - k)$ to $\phi(x_2, \gamma_0 + k)$ and therefore passes once and but once through every intermediate value. Hence if γ_2 be any value of γ in $(\gamma_0 - k, \gamma_0 + k)$ and we put $\phi(x_2, \gamma_2) = y_2$, then the equation $y_2 = \phi(x_2, \gamma)$ has in $(\gamma_0 - k, \gamma_0 + k)$ no other solution but $\gamma = \gamma_2$, which means geometrically that through the point (x_2, y_2)—which is any point of $\mathfrak{S}_k$—there passes but one extremal of the set (8) for which $|\gamma - \gamma_0| \leq k$.

The existence of the single-valued function $\gamma = \psi(x, y)$ being thus established, the existence and continuity of its first partial derivatives follows from the theorem[3] on implicit functions, since

$$\phi_\gamma(x, \gamma) \neq 0 \quad \text{in } \mathfrak{B}_k .$$

[1] Another proof is given by OSGOOD, *loc. cit.*, p. 113.

[2] Viz., the theorems on "uniform continuity" and on the existence of a minimum. Compare E. II A, pp. 18, 19, 49; J. I, Nos. 62, 63, 64, and P., Nos. 19 VI, VII, and 100 VI, VII.

[3] See p. 35, footnote 2.

The values of these partial derivatives are obtained from (9) by the ordinary rules for the differentiation of implicit functions:

$$\gamma_x = -\frac{\phi_x}{\phi_\gamma}, \qquad \gamma_y = \frac{1}{\phi_\gamma} . \tag{12}$$

In case the function $\phi(x, \gamma)$ is regular in $\mathfrak{B}_k$, also the function $\psi(x, y)$ will be regular in $\mathfrak{S}_k$; compare E. II B, p. 103, and HARKNESS AND MORLEY, *Introduction to the Theory of Analytic Functions*, No. 156.

At the same time we see that the set of points $\mathfrak{S}_k$ is identical with the strip of the x, y-plane bounded by the two non-intersecting curves

$$y = \phi(x, \gamma_0 - k) \quad \text{and} \quad y = \phi(x, \gamma_0 + k)$$

on the one hand, and the two lines $x = x_0$ and $x = x_1$ on the other hand.

Finally, a neighborhood (ρ) of the arc $\mathfrak{E}_0$ can be assigned which is wholly contained in $\mathfrak{S}_k$.

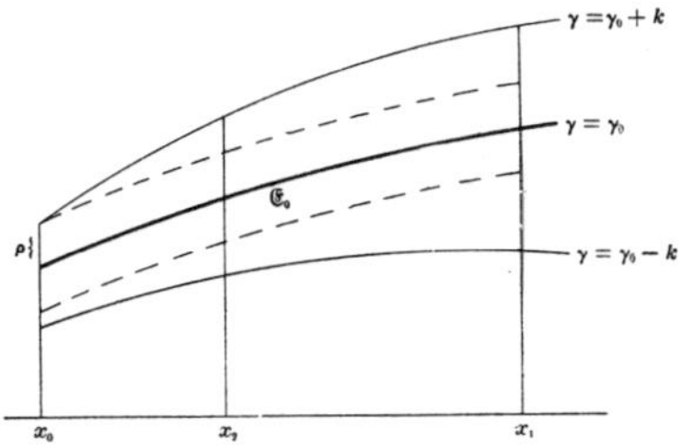

FIG. 17

For each of the two continuous functions $\phi(x, \gamma_0 + k) - \phi(x, \gamma_0)$ and $\phi(x, \gamma_0) - \phi(x, \gamma_0 - k)$ has a positive minimum value in $(x_0 x_1)$; hence if ρ be the smaller of these two minimum values, the neighborhood (ρ) of $\mathfrak{E}_0$ is entirely contained in $\mathfrak{S}_k$.

The region $\mathfrak{S}_k$ has therefore the three characteristic properties of a "field," and the above theorem is proved. (From what has been proved in the first paragraph above, it follows that $\mathfrak{S}_k$ is indeed a region in the specific sense of §2, a).

Corollary I: The slope at a point (x, y) *of the unique extremal of the field* passing through (x, y) is likewise a single-valued function of x, y, which we denote by $p(x, y)$. It is defined analytically by the two equations

$$p(x, y) = \phi_x(x, \gamma), \qquad \gamma = \psi(x, y), \tag{13}$$

which show at the same time that $p(x, y)$ has continuous first partial derivatives in $\mathfrak{S}_k$.

In case $\phi(x, \gamma)$ is regular in $\mathfrak{B}_k$, also $p(x, y)$ is regular in $\mathfrak{S}_k$.

Corollary II: The slope $p(x, y)$ *satisfies the following partial differential equation of the first order:*[1]

$$(p_x + p p_y) F_{y'y'} + p F_{y'y} + F_{y'x} - F_y = 0, \tag{14}$$

the arguments of the partial derivatives of F *being* $x, y, p(x, y)$.

[1] This corollary forms part of HILBERT's proof of Weierstrass's theorem; see below, §21, and the references there given.

Proof: From (13) we obtain by differentiation

$$p_x = \phi_{xx} + \phi_{x\gamma}\gamma_x \ , \quad p_y = \phi_{x\gamma}\gamma_y \ ;$$

hence if we make use of (12) we get

$$p_x + pp_y = \phi_{xx} \ .$$

But since $\phi(x, \gamma)$ satisfies Euler's equation for every value of γ, we have, for every value of x and γ,

$$\phi_{xx}F_{y'y'} + \phi_x F_{y'y} + F_{y'x} - F_y = 0 \ ,$$

the arguments of the partial derivatives of F being x, $\phi(x, \gamma)$, $\phi_x(x, \gamma)$. Hence, if we express γ in terms of x, y by means of (9), we obtain (14).

c) Application to the set of extremals through the point[1] $\bar{A}$.

We can now establish the following

Theorem: If for the extremal $\mathfrak{E}_0$ *the conditions*

$$R > 0 \ , \tag{II'}$$

$$\Delta(x, x_0) \neq 0 \quad \text{for} \quad x_0 < x \leqq x_1 \tag{III'}$$

are fulfilled, and if a point $\bar{A}$ *be chosen on the continuation*[2] *of* $\mathfrak{E}_0$ *beyond* A, *but sufficiently near to* A, *then the set of extremals through* $\bar{A}$ *furnishes a field about* $\mathfrak{E}_0$.

It is only necessary to choose the point $\bar{A}$ (x_5, y_5) so near to A that

1. $X_0 < x_5 < x_0$,
2. $\Delta(x, x_5) \neq 0$ in $(x_0 x_1)$.

The possibility of such a choice of x_5 has been established in §14.

Under these circumstances, it follows by the method employed in §15 that there exists a set of extremals through $\bar{A}$:

$$y = \phi(x, \gamma) \ , \tag{15}$$

[1] The introduction of the set of extremals through $\bar{A}$ instead of the set through A, which considerably simplifies the proofs, is due to ZERMELO, *Dissertation*, pp. 87, 88; compare also KNESER, *Lehrbuch*, §§14, 17 and OSGOOD, *loc. cit.*, p. 115.

[2] Compare the assumptions in §13 *a*).

where[1] $\phi(x, \gamma)$, its first partial derivatives and the derivatives ϕ_{xx}, $\phi_{x\gamma}$ are continuous in the domain

$$X_0 \leqq x \leqq X_1 , \qquad |\gamma - \gamma_0| \leqq d_0 ,$$

d_0 being a sufficiently small positive quantity.

Moreover

$$\phi_\gamma(x, \gamma_0) \neq 0 \qquad \text{in } (x_0 x_1) ,$$

since, corresponding to equation (30) of §15, we have in the present case

$$\phi_\gamma(x, \gamma_0) = C \,.\, \Delta(x, x_5) ,$$

where C is a constant different from zero.

Hence the set of extremals through[2] $\bar{A}$ satisfies the conditions of the lemma given under *b*) and furnishes therefore indeed a field about $\mathfrak{E}_0$.

[1] Notice that in §15 the symbol $\phi(x, \gamma)$ was used with a slightly different meaning, viz., for the set of extremals through A.

[2] To the *set of extremals through the point A itself* the lemma cannot be applied, since for this set $\phi_\gamma(x_0, \gamma_0) = 0$. Nevertheless it can be proved that in this case through every point of $\mathfrak{S}_k$, *except the point A itself*, a unique extremal of the set can be drawn. For in the present case we have: $\phi(x_0, \gamma) \equiv y_0$ for every γ and therefore $\phi_\gamma(x_0, \gamma) \equiv 0$. Hence it follows that if we define

$$\chi(x, \gamma) = \begin{cases} \phi_\gamma(x, \gamma)/(x - x_0) , & \text{when } x \neq x_0 , \\ \phi_{\gamma x}(x_0, \gamma) , & \text{when } x = x_0 , \end{cases}$$

the function $\chi(x, \gamma)$ is continuous in the domain: $X_0 \leqq x \leqq X_1$, $|\gamma - \gamma_0| \leqq d_0$, and $\chi(x, \gamma_0) \neq 0$ in $(x_0 x_1)$, also for $x = x_0$, since $\phi_{\gamma x}(x_0, \gamma_0) \neq 0$ according to equation (31) of §15. We can therefore take k so small that $\chi(x, \gamma) \neq 0$ in the domain: $x_0 \leqq x \leqq x_1$, $|\gamma - \gamma_0| \leqq k$. Hence it follows that $\phi_\gamma(x, \gamma)$ has the same sign throughout the domain: $x_0 < x \leqq x_1$, $|\gamma - \gamma_0| \leqq k$. The further reasoning proceeds then as under *b*).

It should also be noticed that in the present case it is impossible to inscribe in $\mathfrak{S}_k$ a neighborhood (ρ) of $\mathfrak{E}_0$, since the width of $\mathfrak{S}_k$ approaches zero as x approaches x_0.

We shall say that the set of extremals through A forms an *improper field* about $\mathfrak{E}_0$.

The inverse function $\psi(x, y)$ and the slope $p(x, y)$ are in this case single-valued and of class C' in $\mathfrak{S}_k$ except at the point (x_0, y_0) where they are indeterminate. But if the point (x, y) approaches the point (x_0, y_0) along a curve $\mathfrak{C}$ of class C' lying entirely in $\mathfrak{S}_k$, then both functions approach determinate finite limiting values. The limit of $\psi(x, y)$ is the parameter γ of that extremal of the set which is tangent to $\mathfrak{C}$ at (x_0, y_0); the limit of $p(x, y)$ is the slope of the curve $\mathfrak{C}$ at (x_0, y_0).

d) The Field-Integral for the set of extremals through the point $\bar{A}$.

Let $P(x_2, y_2)$ be any point in the field $\mathfrak{F}_k$ formed by the set of extremals through the point $\bar{A}(x_5, y_5)$, and let $\gamma_2 = \psi(x_2, y_2)$ be the value of γ for the unique extremal of the field which passes through the point P. Then the integral J taken along this extremal

$$\mathfrak{E}_2: \qquad y = \phi(x, \gamma_2)$$

from the point $\bar{A}$ to the point P is a single-valued function of x_2, y_2 which we denote by $J(x_2, y_2)$. Its value is

$$J(x_2, y_2) = \int_{x_5}^{x_2} F\big(x, \phi(x, \gamma_2), \phi_x(x, \gamma_2)\big)\, dx ,$$

where it is understood that γ_2 is replaced by its expression $\psi(x_2, y_2)$ in terms of x_2 and y_2.

The partial derivatives of $J(x_2, y_2)$ with respect to x_2 and y_2 have the following values:

$$\frac{\partial J(x_2, y_2)}{\partial x_2} = F(x_2, y_2, p_2) - p_2 F_{y'}(x_2, y_2, p_2) ,$$

$$\frac{\partial J(x_2, y_2)}{\partial y_2} = F_{y'}(x_2, y_2, p_2) , \tag{15a}$$

where p_2 denotes the slope of the extremal $\mathfrak{E}_2$ at the point P.

For

$$\frac{\partial J(x_2, y_2)}{\partial x_2} = F(x_2, y_2, p_2) + \frac{\partial \gamma_2}{\partial x_2} \int_{x_5}^{x_2} (F_y \phi_\gamma + F_{y'} \phi_{x\gamma})\, dx ,$$

$$\frac{\partial J(x_2, y_2)}{\partial y_2} = \frac{\partial \gamma_2}{\partial y_2} \int_{x_5}^{x_2} (F_y \phi_\gamma + F_{y'} \phi_{x\gamma})\, dx .$$

If we transform the integral as in §20, c), and make use of (12) we obtain (15a).

In many respects it would have been preferable first to prove the formulæ (15a) and to make use of them in the demonstration of WEIERSTRASS's theorem.

Compare the analogous formulæ (44) in §37, and the still more general formulæ (14) in §34.

§20. WEIERSTRASS'S THEOREM

We are now prepared to prove a fundamental theorem whose discovery by WEIERSTRASS in 1879 marks a turning-point in the history of the Calculus of Variations. It gives an expression for the total variation of the integral J in terms of the **E**-function, from which sufficient conditions for an extremum can be derived.

a) The gist of Weierstrass's method can be best understood from a simple example, in which the difficulties concerning the existence of a field, which complicate the proof of Weierstrass's theorem in the general case, can be entirely avoided.

EXAMPLE VI: In order to prove that the straight line[1] 01 actually minimizes the integral

$$J = \int_{x_0}^{x_1} \sqrt{1 + y'^2}\, dx \;,$$

we draw from the point 0 to the point 1 any curve $\bar{\mathfrak{C}}$:

$$\bar{\mathfrak{C}}: \qquad \bar{y} = \bar{f}(x) \;,$$

[1] For the notation compare §2, e).

not coinciding with the straight line 01. We suppose for simplicity that $\overline{\mathfrak{C}}$ is of class C'.

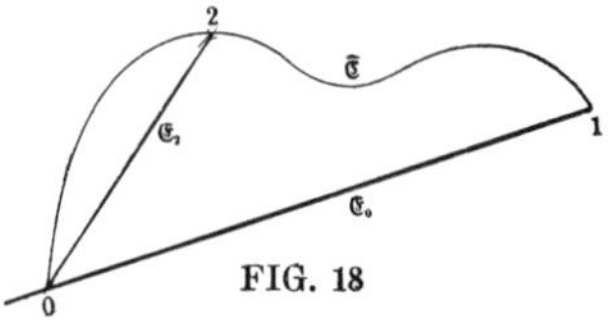

FIG. 18

Through an arbitrary point $2:(x_2, y_2)$ of $\overline{\mathfrak{C}}$ we can draw one and but one extremal of the set of extremals through the point 0, viz., the straight line 02.

We now consider the integral J taken from 0 along the straight line 02 to 2 and from 2 along the curve $\overline{\mathfrak{C}}$ to 1, that is, we form, in the notation of §2, f),

$$J_{02} + \bar{J}_{21} ,$$

the stroke always indicating[1] integration along the curve $\overline{\mathfrak{C}}$.

The value of this integral is a single-valued function of x_2, which will be denoted by $S(x_2)$, As the point 2 describes the curve $\overline{\mathfrak{C}}$ from 0 to 1, $S(x_2)$ varies continuously[2] from the initial value

$$S(x_0) = \bar{J}_{01} \quad (\text{along } \overline{\mathfrak{C}})$$

to the end value

$$S(x_1) = J_{01} \quad (\text{along } \mathfrak{C}_0) .$$

Hence the total variation

$$\Delta J = \bar{J}_{01} - J_{01}$$

is expressible in terms of the function $S(x)$ in the form

[1] Notation according to WEIERSTRASS; KNESER, on the contrary, uses the stroke to indicate integration along an extremal.

[2] See the explicit expressions for J_{02} and $\bar{J}_{21}$ below.

$$\Delta J = -\left[S(x_1) - S(x_0)\right] ;$$

and we shall have proved that $\Delta J \geqq 0$ if we can show that $S(x_2)$ always decreases or at least does not increase as x_2 increases from x_0 to x_1.

For this purpose we form the derivative of $S(x_2)$.

The integral J_{02} is the length of the straight line 02:

$$J_{02} = \sqrt{(x_2 - x_0)^2 + (y_2 - y_0)^2} ;$$

hence

$$\frac{dJ_{02}}{dx_2} = \frac{(x_2 - x_0) + (y_2 - y_0)\bar{f}'(x_2)}{\sqrt{(x_2 - x_0)^2 + (y_2 - y_0)^2}} ,$$

since

$$y_2 = \bar{f}(x_2) .$$

If we denote the slopes of the straight line 02 and of the curve $\bar{\mathfrak{C}}$ at 2 respectively by p_2 and $\bar{p}_2$, *i. e.*,

$$p_2 = \frac{y_2 - y_0}{x_2 - x_0} , \qquad \bar{p}_2 = \bar{f}'(x_2) ,$$

the previous result may be written

$$\frac{dJ_{02}}{dx_2} = \frac{1 + p_2\bar{p}_2}{\sqrt{1 + p_2^2}} .$$

On the other hand,

$$\bar{J}_{21} = \int_{x_2}^{x_1} \sqrt{1 + \bar{f}'^2(x)}\, dx ,$$

and therefore
$$\frac{d\bar{J}_{21}}{dx_2} = -\sqrt{1+\bar{p}_2^2}\,.$$

Hence we obtain the result

$$\frac{dS(x_2)}{dx_2} = -\sqrt{1+\bar{p}_2^2}\left\{1 - \frac{1+p_2\bar{p}_2}{\sqrt{1+p_2^2}\sqrt{1+\bar{p}_2^2}}\right\},$$

from which we easily infer that

$$\frac{dS(x_2)}{dx_2}\begin{cases} < 0 & \text{if } \bar{p}_2 \neq p_2\,, \\ = 0 & \text{if } \bar{p}_2 = p_2\,. \end{cases}$$

The latter alternative cannot take place[1] all along the curve $\bar{\mathfrak{C}}$. Hence it follows that

$$\Delta J > 0\,.$$

The reasoning can easily be extended to the case in which the curve $\bar{\mathfrak{C}}$ has a finite number of corners.

It is thus proved that *the straight line* 01 *furnishes a proper*[2] *absolute*[2] *minimum* for the integral J.

The preceding construction may be modified[3] as follows: On the continuation of the line $\mathfrak{C}_0$ beyond the point 0 choose a point 5, and replace in the preceding construction the line 02 by the line 52. Accordingly the function $S(x_2)$ is now defined by:

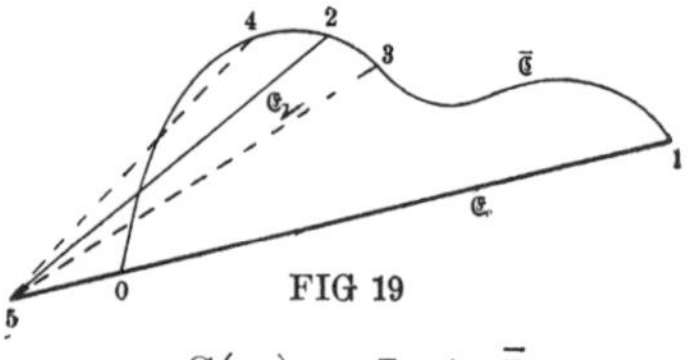

FIG 19

$$S(x_2) = J_{52} + \bar{J}_{21}\,,$$

and therefore

$$S(x_0) = J_{50} + \bar{J}_{01}\,, \qquad S(x_1) = J_{51} = J_{50} + J_{01}\,.$$

Hence we have again

$$\Delta J = -\left[S(x_1) - S(x_0)\right].$$

[1] If $\bar{p}_2 = p_2$ for every x_2 in (x_0x_1) it would follow that $\bar{f}(x)$ satisfies the differential equation

$$\bar{f}(x) - y_0 = (x - x_0)\bar{f}'(x)\,,$$

and therefore $\bar{\mathfrak{C}}$ must be a straight line through 0, which could be no other than the line $\mathfrak{C}_0$, since $\bar{\mathfrak{C}}$ is to pass through 1.

[2] Compare §3, *a*) and *b*).

[3] Compare p. 82, footnote 1.

For the derivative of $S(x_2)$ we obtain the same expression as before, if we let, in the present case, p_2 denote the slope of the extremal 52.

b) We now proceed to the *general case.* We suppose that for the extremal $\mathfrak{E}_0$ the conditions (II′) and (III′) are fulfilled. Then we construct as in §19, *c*) a field $\mathfrak{F}_k$ about $\mathfrak{E}_0$ by means of the set of extremals (15) through the point $\bar{A}$, chosen as indicated in §19, *c*) on the continuation of $\mathfrak{E}_0$ beyond A. Since the extremal $\mathfrak{E}_0$ is supposed to lie in the interior[1] of the region $\mathfrak{R}$, we can take k so small that $\mathfrak{F}_k$ is entirely contained in $\mathfrak{R}$.

For our present purpose it will be convenient to use the numbers 0, 1, 5 to denote the points A, B, $\bar{A}$ respectively. Let now $\bar{\mathfrak{C}}$ be any curve of class C' joining the two points 0 and 1 (see Fig. 19), and lying wholly in the field $\mathfrak{F}_k$, and let 2 be an arbitrary point of $\bar{\mathfrak{C}}$. Through the point 2 we can draw one and but one extremal of the field, *i. e.*, one extremal of the set (15) for which $|\gamma-\gamma_0|\leqq k$; let it be denoted by

$$\mathfrak{E}_2: \qquad y=\phi(x,\gamma_2)\,.$$

We then consider the integral J taken from 5 to 2 along $\mathfrak{E}_2$ and from 2 to 1 along $\bar{\mathfrak{C}}$, and denote its value by $S(x_2)$:

$$S(x_2)=J_{52}+\bar{J}_{21}=\int_{x_5}^{x_2}F\,dx+\int_{x_2}^{x_1}\bar{F}\,dx\,, \tag{16}$$

the arguments of F being

$$x\,,\qquad y=\phi(x,\gamma_2)\,,\qquad y'=\phi_x(x,\gamma_2)\,,$$

those of $\bar{F}$:

$$x\,,\qquad \bar{y}=\bar{f}(x)\,,\qquad \bar{y}'=\bar{f}'(x)\,.$$

For $x_2=x_0$ and $x_2=x_1$, $S(x_2)$ takes the values[2]

$$S(x_0)=J_{50}+\bar{J}_{01}\,,\qquad S(x_1)=J_{51}\,, \tag{17}$$

[1] See §11.

[2] Properly speaking, $S(x_2)$ is not defined for $x_2=x_1$. But in order that $S(x_2)$ may be continuous also at $x_2=x_1$, we must define $S(x_1)=S(x_1-0)$; and $S(x_1-0)$ is easily seen to be equal to J_{51}.

so that $$\Delta J = \bar{J}_{01} - J_{01} = -\left[S(x_1) - S(x_0)\right] . \tag{18}$$

The function $S(x_2)$ *is continuous and admits in* (x_0x_1) *a derivative whose value is*

$$S'(x_2) = -\mathbf{E}(x_2, y_2; p_2, \bar{p}_2) , \tag{19}$$

where $\bar{p}_2$ denotes the slope of $\bar{\mathfrak{C}}$, p_2 that of $\mathfrak{E}_2$, at the point 2.

WEIERSTRASS[1] reaches these results in the following way:

Let 3 denote that point of $\bar{\mathfrak{C}}$ whose abscissa is $x_2 + h$, h being a small positive quantity; and let

$$\mathfrak{E}_3: \qquad y = \phi(x, \gamma_2 + \epsilon)$$

be the unique extremal of the field which passes through the point 3. Then

$$S(x_2 + h) - S(x_2) = (J_{53} + \bar{J}_{31}) - (J_{52} + \bar{J}_{21}) = J_{53} - (J_{52} + \bar{J}_{23}) .$$

But this is precisely the difference which has been computed[2] in §8, equation (30), the curves $\mathfrak{E}_2$, $\mathfrak{E}_3$, $\bar{\mathfrak{C}}$ corresponding to the curves there denoted by $\mathfrak{C}$, $\bar{\mathfrak{C}}$, $\tilde{\mathfrak{C}}$. Accordingly we obtain

$$S(x_2 + h) - S(x_2) = -h\left[\mathbf{E}(x_2, y_2; p_2, \bar{p}_2) + (h)\right] , \tag{20}$$

(h) denoting an infinitesimal.

Similarly, if 4 be that point of $\bar{\mathfrak{C}}$ whose abscissa is $x_2 - h$, we obtain

$$S(x_2 - h) - S(x_2) = J_{54} + \bar{J}_{42} - J_{52} ,$$

which, according to the lemma of §8, is equal to

$$S(x_2 - h) - S(x_2) = +h\left[\mathbf{E}(x_2, y_2; p_2, \bar{p}_2) + (h)\right] . \tag{20a}$$

Hence the derivative of S exists and its value is indeed given by (19).

[1] *Lectures*, 1879; the proof here given is Weierstrass's original proof with the necessary adaptations to the case where x is the independent variable, and with the substitution of the set of extremals through 5 for the set through 0.

[2] In applying the lemma of §8 to the present case, we have to make use of the remarks on p. 18 and p. 35. The variation

$$\Delta y = \phi(x, \gamma_2 + \epsilon) - \phi(x, \gamma_2)$$

is indeed a variation of the type (5a) of §4, *d*).

As the point 2 describes the curve $\bar{\mathfrak{C}}$ from 0 to 1, the function $\mathbf{E}(x_2, y_2; p_2, \bar{p}_2)$ varies continuously. For, on the one hand the E-function is a continuous function of its four arguments, provided that the point (x, y) remains in the region $\mathfrak{R}$, and the field $\mathfrak{S}_k$ is contained in $\mathfrak{R}$; on the other hand, $y_2 = \bar{f}(x_2)$ and $\bar{p}_2 = \bar{f}'(x_2)$ are continuous in $(x_0 x_1)$ and the slope p_2 of $\mathfrak{C}_2$ at 2 is, according to §19, *b*), a continuous function of x_2, y_2.

Integrating (19) between the limits x_0 and x_1, and remembering (18), we obtain therefore *for the total variation* ΔJ *the expression*[1]

$$\Delta J = \int_{x_0}^{x_1} \mathbf{E}(x_2, y_2; p_2, \bar{p}_2)\, dx_2 \,. \tag{21}$$

We shall refer to this important formula as "WEIERSTRASS'S *theorem.*"

The theorem remains true for curves $\bar{\mathfrak{C}}$ of class D'. For, suppose the curve $\bar{\mathfrak{C}}$ to have a corner at the point 2. Then (20) and (20a) still hold if we understand by $\bar{p}_2$ the progressive and regressive derivatives of $\bar{f}(x_2)$ respectively. The function $S(x)$ is therefore continuous at x_2 and admits a progressive and a regressive derivative. Hence it follows[2] that (21) still holds when $\bar{\mathfrak{C}}$ has a finite number of corners.

c) Instead of first computing the increments $S(x_2 \pm h) - S(x_2)$, KNESER (*Lehrbuch*, §20) and OSGOOD (*loc. cit.*, p. 116) compute directly the derivative $S'(x_2)$ by applying the theorem on *the differentiation of a definite integral with respect to a parameter.* Supposing for simplicity that $\bar{\mathfrak{C}}$ is of class C', it follows from the properties of the function $\phi(x, \gamma)$ that $S(x_2)$ is continuous and differentiable in the

[1] The theorem remains true also for the "improper field" $\mathfrak{S}_k$ formed by the set of extremals through the point 0, and for a curve $\bar{\mathfrak{C}}$ which lies entirely in this field $\mathfrak{S}_k$. For formula (19) holds also in this case at every point of $\bar{\mathfrak{C}}$ with the exception of the point 0. Integrating (19) from $x_0 + h$ to x_1, and passing to the limit $h = 0$, we obtain (21) since p_2 approaches a determinate finite limit; compare footnote 2, p. 83.

[2] Compare E. II A, p. 100, and DINI, §194.

interval $(x_0 x_1)$ and that the derivative can be obtained by applying to the definite integrals J_{52} and $\bar{J}_{21}$ the ordinary rules[1] for the differentiation of a definite integral with respect to a parameter and with respect to the limits.

Accordingly we obtain in the first place

$$\frac{d\bar{J}_{21}}{dx_2} = -F(x_2, y_2, \bar{p}_2) \ . \tag{22}$$

In differentiating the integral

$$J_{52} = \int_{x_5}^{x_2} F\big(x, \phi(x, \gamma_2), \phi_x(x, \gamma_2)\big) dx \ ,$$

we must remember that γ_2 is a function of x_2 defined by the equation

$$\phi(x_2, \gamma_2) = \bar{f}(x_2) \ , \tag{23}$$

which expresses the fact that the curves $\mathfrak{E}_2$ and $\bar{\mathfrak{C}}$ both pass through the point 2.

Accordingly we obtain:

$$\frac{dJ_{52}}{dx_2} = F(x_2, y_2, p_2) + \int_{x_5}^{x_2} (F_y \phi_\gamma + F_{y'} \phi_{x\gamma}) \frac{d\gamma_2}{dx_2} dx \ ,$$

the arguments of ϕ_γ, $\phi_{x\gamma}$ being x, γ_2.

From our assumptions concerning $\phi(x, \gamma)$ it follows that

$$\phi_{x\gamma}(x, \gamma_2) = \phi_{\gamma x}(x, \gamma_2) \ .$$

Applying then to the second term under the integral sign the integration by parts of §4, and remembering that the function $y = \phi(x, \gamma_2)$ is an integral of Euler's differential equation

$$F_y - \frac{d}{dx} F_{y'} = 0 \ ,$$

we obtain the result:

$$\frac{dJ_{52}}{dx_2} = F(x_2, y_2, p_2) + \frac{d\gamma_2}{dx_2} \Big[F_{y'}(x_2, y_2, p_2) \phi_\gamma(x_2, \gamma_2) \\ - F_{y'}(x_5, y_5, p_5) \phi_\gamma(x_5, \gamma_2) \Big] \ ,$$

where $p_5 = \phi_x(x_5, \gamma_2)$.

[1] Compare E. II A, p. 102, and J. I, No. 83.

But since the extremals of the set (15) all pass through the point $5 : (x_5, y_5)$, we have

$$y_5 = \phi(x_5, \gamma)$$

for every γ; hence

$$\phi_\gamma(x_5, \gamma) = 0$$

for every γ, and therefore in particular

$$\phi_\gamma(x_5, \gamma_2) = 0 \ .$$

On the other hand, if we differentiate (23) with respect to x_2, we get

$$\phi_\gamma(x_2, \gamma_2) \frac{d\gamma_2}{dx_2} = \bar{p}_2 - p_2 \ ;$$

therefore

$$\frac{dJ_{52}}{dx_2} = F(x_2, y_2, p_2) + (\bar{p}_2 - p_2) F_{y'}(x_2, y_2, p_2) \ . \tag{24}$$

Combining (22) and (24) we obtain again the fundamental formula (19).

§21. HILBERT'S PROOF OF WEIERSTRASS'S THEOREM

Weierstrass's theorem can be extended[1] to *any* set of extremals constituting a field about the arc $\mathfrak{E}_0$, *i. e.*,

Whenever the extremal $\mathfrak{E}_0$ can be surrounded by a field, the total variation $\Delta J = J_{\bar{\mathfrak{E}}} - J_{\mathfrak{E}_0}$ for any admissible curve $\bar{\mathfrak{E}}$ lying wholly in the field, is expressible by WEIERSTRASS'S *formula:*

$$\Delta J = \int_{x_0}^{x_1} \mathbf{E}(x, \bar{y}; p, \bar{p})\, dx \ ,$$

where $(x, \bar{y})$ is a point of $\bar{\mathfrak{E}}$, $\bar{p}$ the slope of $\bar{\mathfrak{E}}$ at $(x, \bar{y})$, and p the slope at $(x, \bar{y})$ of the unique extremal of the field passing through $(x, \bar{y})$.

[1] The extension seems to be due to H. A. SCHWARZ, who has given the generalized theorem in a course of lectures in 1898–99.

The following elegant proof of the generalized theorem is due to HILBERT.[1]

Suppose $\mathfrak{S}_k$ is a field of extremals about our extremal $\mathfrak{E}_0$. In $\mathfrak{S}_k$ we draw any curve $\mathfrak{C}$ of class D': $y=f(x)$, joining A and B. Now let $p(x, y)$ be an arbitrary function of x, y which is of class C' in $\mathfrak{S}_k$, and consider the integral

$$J^* = \int_{x_0}^{x_1} \Big[F\big(x, y, p(x, y)\big) \\ + \big(y' - p(x, y)\big) F_{y'}\big(x, y, p(x, y)\big)\Big]\, dx \tag{25}$$

taken along the curve $\mathfrak{C}$ from A to B. The value of J^* will, in general, depend upon the choice of the curve $\mathfrak{C}$; we ask: How must we choose the function $p(x, y)$ in order that the value of J^* may be independent of the choice of the curve $\mathfrak{C}$ and dependent only upon the position of the two end-points A and B?

Our integral J^* is of the form

$$\int_{x_0}^{x_1} \big[M(x, y) + N(x, y)\, y'\big]\, dx\ ,$$

and it has been seen in §7, d) that the necessary and sufficient[2] condition that such an integral should be independent of the path of integration is that

$$M_y \equiv N_x\ .$$

In the present case we have

$$M(x, y) = F(x, y, p) - p\, F_{y'}(x, y, p)\ ,$$
$$N(x, y) = F_{y'}(x, y, p)\ ;$$

hence

$$M_y = F_y - p\,(F_{y'y} + p_y F_{y'y'})\ ,$$
$$N_x = F_{y'x} + p_x F_{y'y'}\ .$$

[1] See *Göttinger Nachrichten*, 1900, pp. 253–297, and *Archiv der Mathematik und Physik* (3), Vol. I (1901), p. 231, *Abh.*, Vol. III, p. 324; also the English translation, in the *Bulletin of the American Mathematical Society* (2), Vol. VIII (1902), p. 473; further, OSGOOD's presentation in the *Annals of Mathematics* (2), Vol. II (1901), p. 121, and HEDRICK, *Bulletin of the American Mathematical Society* (2), Vol. IX (1902), p. 11.

[2] Notice that the region $\mathfrak{S}_k$, to which the curves $\mathfrak{C}$ are confined, is simply connected.

Hence, *in order that the value of the integral J^* may be independent of the path of integration $\mathfrak{C}$, it is necessary and sufficient that the function $p(x, y)$ satisfy the partial differential equation*

$$(p_x + pp_y) F_{y'y'} + p F_{y'y} + F_{y'x} - F_y = 0 \ , \tag{26}$$

the arguments of the partial derivatives of F being $x, y, p(x, y)$.

But this differential equation is identical with the differential equation (14) which is satisfied by the slope at (x, y) of the extremal of the field passing through (x, y). Hence the value of J^* will be independent of the choice of the curve $\mathfrak{C}$, if we select for the function p the slope just defined. In the sequel p will have this special meaning.

The invariance of the integral J^* being established, we select for the curve $\mathfrak{C}$ first the extremal $\mathfrak{E}_0$; then we have all along $\mathfrak{E}_0$:

$$y' = p(x, y) \ ,$$

because $\mathfrak{E}_0$ is the unique extremal of the field which passes through a point of $\mathfrak{E}_0$. Therefore (25) reduces to

$$J^* = \int_{x_0}^{x_1} F(x, y, y')\,dx = J_{\mathfrak{E}_0} \ .$$

On the other hand, if we select for $\mathfrak{C}$ any curve $\overline{\mathfrak{C}}$ of class D', different from $\mathfrak{E}_0$, and joining A and B, we get

$$J^* = \int_{x_0}^{x_1} \left[F(x, \bar{y}, p) + (\bar{p} - p) F_{y'}(x, \bar{y}, p) \right] dx \ ,$$

where $\bar{p} = \bar{y}'$ denotes the slope of $\overline{\mathfrak{C}}$ at the point $(x, \bar{y})$. Both values of J^* being equal on account of the invariance of J^*, we obtain an expression for $J_{\mathfrak{E}_0}$ in terms of a definite integral taken along $\mathfrak{C}$. This expression we use in forming the total variation

$$\Delta J = J_{\overline{\mathfrak{C}}} - J_{\mathfrak{E}_0} \ .$$

Then we obtain

$$\Delta J = \int_{x_0}^{x_1} \Big[F(x, \bar{y}, \bar{p}) - F(x, \bar{y}, p) - (\bar{p} - p) F_{y'}(x, \bar{y}, p) \Big] \, dx \, ,$$

which is the desired extension of WEIERSTRASS's theorem, since the integrand is equal to $\mathbf{E}(x, \bar{y}; \ p, \bar{p})$.

§22. SUFFICIENT CONDITIONS FOR A STRONG MINIMUM[1]

Weierstrass's theorem leads now immediately to sufficient conditions for a strong minimum:

a) Suppose there exists a field $\mathfrak{S}_k$ about $\mathfrak{E}_0$ such that at every point of $\mathfrak{S}_k$

$$\mathbf{E}\big(x, y; \, p(x, y), \tilde{p}\big) \geqq 0 \tag{27}$$

for every finite value of $\tilde{p}$, $p(x, y)$ denoting again the slope at (x, y) of the extremal of the field passing through (x, y).

Then it follows from Weierstrass's theorem that $\Delta J \geqq 0$ for every curve $\bar{\mathfrak{C}}$ of class D' drawn in $\mathfrak{S}_k$ from A to B, and moreover that $\Delta J > 0$ unless

$$\mathbf{E}\big(x, \bar{y}; \, p(x, \bar{y}), \bar{y}'\big) = 0 \tag{28}$$

all along the curve $\bar{\mathfrak{C}}$.

From the definition of the E-function it follows that (28) holds at a point $(x, \bar{y})$ of $\bar{\mathfrak{C}}$ whenever

$$\bar{y}' = p(x, \bar{y}) \, ,$$

i. e., whenever the extremal through $(x, \bar{y})$ is tangent to $\bar{\mathfrak{C}}$ at $(x, \bar{y})$. This can, however, not take place at every point of $\bar{\mathfrak{C}}$, unless $\bar{\mathfrak{C}}$ completely coincides with $\mathfrak{E}_0$. For[2] the value of the parameter γ of the extremal of the field passing through that point of $\bar{\mathfrak{C}}$ whose abscissa is x, is determined by the equation

$$\bar{f}(x) = \phi(x, \gamma) \, ,$$

[1] Compare for this section also HEDRICK, *Bulletin of the American Mathematical Society*, Vol. IX (1902), p. 11.

[2] This proof is due to KNESER, *Lehrbuch*, §22; see also OSGOOD, *loc. cit.*, p. 118.

from which we derive by differentiation

$$\bar{f}'(x) = \phi_x(x, \gamma) + \phi_\gamma(x, \gamma)\frac{d\gamma}{dx},$$

or according to (13)

$$\bar{y}' - p(x, \bar{y}) = \phi_\gamma(x, \gamma)\frac{d\gamma}{dx}.$$

But according to (11), $\phi_\gamma(x, \gamma) \neq 0$; if therefore $\bar{y}' = p(x, \bar{y})$ at every point of $\bar{\mathfrak{C}}$, we should have

$$\frac{d\gamma}{dx} = 0 \qquad \text{throughout } (x_0 x_1),$$

or $\gamma = \text{const.}$, *i. e.*, $\bar{\mathfrak{C}}$ would itself be an extremal of the field, which could be no other than $\mathfrak{C}_0$, since $\bar{\mathfrak{C}}$ passes through the point (x_1, y_1) and $\mathfrak{C}_0$ is the only extremal of the field which passes through (x_1, y_1).

Hence, if instead of (27) the stronger condition[1]

$$\mathbf{E}_1\big(x, y;\, p(x, y), \tilde{p}\big) > 0 \tag{29}$$

is satisfied at every point (x, y) of $\mathfrak{S}_k$ and for every finite $\tilde{p}$, it follows that $\Delta J > 0$ for every admissible curve $\bar{\mathfrak{C}}$ drawn in the field $\mathfrak{S}_k$.

In the terminology of §3 we have therefore the result that *whenever (27) is satisfied, $\mathfrak{C}_0$ furnishes a minimum for the integral J; if moreover (29) is satisfied*[2] *the minimum is a "proper minimum."*

EXAMPLE III (see pp. 73, 78):

$$F = y'^2(y'+1)^2.$$

The set of straight lines

$$y = mx + \gamma$$

parallel to the extremal AB furnishes evidently a field about $\mathfrak{C}_0$, and for this field

$$p(x, y) = m.$$

Therefore

[1] Compare (6) and (6a).

[2] It is even sufficient that (27) and (29) be satisfied in a neighborhood (ρ) of $\mathfrak{C}_0$ inscribed in $\mathfrak{S}_k$; the same remark applies later on to (IIb').

$$\mathbf{E}\left(x, y; p(x, y), \tilde{p}\right) = (\tilde{p} - m)^2 \left[\tilde{p}^2 + 2\tilde{p}(m+1) + 3m^2 + 4m + 1\right].$$

When $m > 0$ or $m < -1$, condition (29) is fulfilled, and therefore the straight line AB actually minimizes the integral

$$J = \int_{x_0}^{x_1} y'^2 (y'+1)^2 \, dx$$

in these two cases.

b) The sufficient conditions thus immediately following from Weierstrass's theorem are, however, in general inconvenient for applications, and it is therefore important to remark that they can be replaced, under certain additional assumptions either concerning the curves $\overline{\mathfrak{C}}$ or concerning the function F, by simpler conditions.

From the relation (5) between the **E**-function and $F_{y'y'}$, it follows that both conditions (27) and (29) are always satisfied when

$$F_{y'y'}(x, y, \tilde{p}) > 0 \qquad \text{(IIb')}$$

at every point[1] (x, y) of $\mathfrak{S}_k$ and for every finite value of $\tilde{p}$.

Hence if we remember the theorem concerning the existence of a field (§19, *b*)), we can state the following theorem:

FUNDAMENTAL THEOREM V:[2] *If the extremal $\mathfrak{E}_0 : AB$ does not contain the conjugate point to A, and if further*

$$F_{y'y'}(x, y, \tilde{p}) > 0 \qquad \text{(IIb')}$$

at every point (x, y) of a certain neighborhood of $\mathfrak{E}_0$ for every finite value of $\tilde{p}$, then $\mathfrak{E}_0$ actually minimizes the integral

$$J = \int_{x_0}^{x_1} F(x, y, y') \, dx .$$

Corollary: The minimum is moreover a "*proper minimum*," *i. e.*, $\Delta J > 0$ for every admissible variation of the curve $\mathfrak{E}_0$ in a certain neighborhood of $\mathfrak{E}_0$.

[1] It is even sufficient that (27) and (29) be satisfied in a neighborhood (ρ) of $\mathfrak{E}_0$ inscribed in $\mathfrak{S}_k$; the same remark applies later on to (IIb').

[2] See OSGOOD, *loc. cit.*, p. 118; compare, however, below, the remark on p. 99, footnote 1.

For a so-called *regular problem* (compare §7, c)) it is therefore sufficient for an extremum that the arc AB does not contain the conjugate to the point A.

EXAMPLE VII:[1]

$$F = g(x, y)\sqrt{1+y'^2}\ ,$$

$g(x, y)$ being a function of x and y alone, of class C'' in a certain region $\mathfrak{R}$. Here

$$F_{y'y'}(x, y, \tilde{p}) = \frac{g(x, y)}{(\sqrt{1+\tilde{p}^2})^3}\ .$$

Hence every extremal AB which lies in the interior of $\mathfrak{R}$ and which does not contain the conjugate point to A, furnishes a minimum provided that $g(x, y) > 0$ along AB. For $g(x, y)$, being continuous in a certain neighborhood of AB and positive along AB, will also be positive in a certain neighborhood of AB, so that (IIb′) is satisfied.

This covers the case of Examples I and VI, in which

$$g(x, y) = y\ , \qquad \text{and } 1 \tag{1}$$

respectively; and also the case of the "brachistochrone" in which

$$g(x, y) = \frac{1}{\sqrt{y - y_0 + k}}\ .$$

All three functions are positive along the respective extremals.

On account of the extension of Weierstrass's theorem given in §21, Theorem V may be replaced by the following:

If the extremal $\mathfrak{E}_0$ can be surrounded by a field and if Condition (IIb′) is fulfilled, then $\mathfrak{E}_0$ actually minimizes the integral J.

Frequently the existence of some particular field about the arc $\mathfrak{E}_0$ is geometrically evident; in such cases the second form of the theorem is more convenient.

[1] Geometrical Interpretation (ERDMANN): Let a straight line move perpendicularly to the x, y-plane along the curve $y = f(x)$ from A to B. The area of that portion of the cylindric surface thus generated which lies between the x, y-plane and the surface: $z = g(x, y)$ is equal to

$$\int_{x_0}^{x_1} g(x, y)\sqrt{1+y'^2}\,dx\ .$$

EXAMPLE VIII:[1] To minimize the integral

$$J = \int_{x_0}^{x_1} \frac{\sqrt{1+y'^2}}{y}\,dx\ ,$$

the admissible curves being confined to the upper half-plane ($y > 0$).

Here the extremals are semi-circles having their centers on the x-axis. If

$$y = \sqrt{-(x-\alpha_0)^2 + \gamma_0^2}$$

is the particular semi-circle passing through the two given points, the set of concentric circles

$$y = \sqrt{-(x-\alpha_0)^2 + \gamma^2} \equiv \phi(x, \gamma)$$

evidently furnishes a field about $\mathfrak{E}_0$. Moreover (IIb′) is fulfilled throughout the upper half-plane. Hence the semi-circle through the two given points actually minimizes the integral J.

Remark: Though the above theorem is the one which is most important for applications, it should be observed that it assumes much more than is necessary. Indeed, the *condition (IIb′) is by no means necessary*, not even the milder condition

$$F_{y'y'}(x, y, \tilde{p}) \geqq 0 \tag{IIa}$$

at every point (x, y) *of* $\mathfrak{E}_0$ and for every finite $\tilde{p}$.

This is illustrated by *Example III* (see pp. 73, 78, 95). For here

$$F_{y'y'}(x, y, \tilde{p}) = 2(6\tilde{p}^2 + 6\tilde{p} + 1)$$

can take negative as well as positive values at every point (x, y), and nevertheless, as we have seen above, a minimum takes place when $m > 0$ or $m < -1$.

c) Question of necessary AND *sufficient conditions.*

From WEIERSTRASS'S results concerning the sufficient conditions for the problem in parameter-representation (see §28), one is led to expect that the conditions[2] (I), (III′),

[1] Given by OSGOOD, *loc. cit.*, pp. 109, 115, where also a geometrical interpretation will be found.

[2] The accent indicates the omission of the equality sign in conditions (III) and (IVa); compare pp. 68, 76. (II′) may be omitted, since it is contained in (IVa′); compare §18, equation (6a).

(IVa′) are sufficient for a minimum. Leaving aside the exceptional case when in one of the inequalities (III), (IVa) the equality sign takes place, we should then have reached a system of necessary and sufficient conditions.

The analogy of the problem in parameter-representation is, however, misleading in this case. As a matter of fact *the three conditions (I), (III′), (IVa′) are* NOT[1] *sufficient for a minimum without some additional assumptions, not even if (IVa′) be replaced by the stronger condition*

$$F_{y'y'}(x, y, \tilde{p}) > 0 \qquad \text{(IIa}')$$

at every point (x, y) *of* $\mathfrak{E}_0$ *for every finite value of* $\tilde{p}$.

To prove this statement it suffices to construct a single example in which the conditions in question are fulfilled and in which, nevertheless, no minimum takes place. Such an example is the following:

EXAMPLE IX:[2] To minimize the integral

$$J = \int_0^1 [ay'^2 - 4byy'^3 + 2bxy'^4]\, dx \ ,$$

a, b being two positive constants, with the initial conditions

$$y = 0 \quad \text{for } x = 0 \ , \qquad \text{and } y = 0 \quad \text{for } x = 1 \ .$$

Here Euler's equation reduces to

$$-y''F_{y'y'} = 0 \ ,$$

where

$$F_{y'y'} \equiv 2a - 24byy' + 24bxy'^2 \ .$$

The only extremal through the two given points $A(0, 0)$ and $B(1, 0)$ is the straight line:

$$\mathfrak{E}_0: \qquad y = 0 \ .$$

[1] This statement seems to contradict directly the theorem given in OSGOOD's article, *loc. cit.*, p. 118. But it is to be remembered that OSGOOD makes (p. 108) the assumption that $F_{y'y'}(x, y, \tilde{p}) \neq 0$ *in a certain neighborhood of* $\mathfrak{E}_0$. This assumption, together with (IIa′), is equivalent to (IIb′).

[2] See BOLZA, "Some Instructive Examples in the Calculus of Variations," *Bulletin of the American Mathematical Society* (2), Vol. IX, p. 9.

The set of extremals through A is the pencil of straight lines through A; hence there exists no conjugate point, and condition (III′) is fulfilled.

Further

$$\mathbf{E}_1(x, y; y', \tilde{p}) \equiv (a - 8byy' + 6bxy'^2) + \tilde{p}(-4by + 4bxy') + 2bx\tilde{p}^2 ;$$

hence along $\mathfrak{E}_0$:

$$\mathbf{E}_1\left(x, f_0(x); f_0'(x), \tilde{p}\right) \equiv a + 2bx\tilde{p}^2 > 0 . \qquad \text{(IVa')}$$

The three conditions (I), (III′), (IVa′) are therefore satisfied, even the stronger condition

$$F_{y'y'}\left(x, f_0(x), \tilde{p}\right) \equiv 2a + 24bx\tilde{p}^2 > 0 . \qquad \text{(IIa')}$$

Nevertheless the line $\mathfrak{E}_0$ does not minimize the integral J.

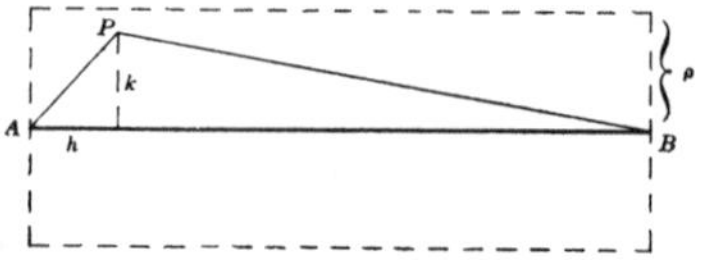

FIG. 20

For, if we replace the line AB by the broken line APB, the co-ordinates of P being $x = h > 0$ and $y = k$, the total variation of J is easily found to be

$$\Delta J = k^2\left[-\frac{bk^2}{h^2} + \frac{a}{h} + a + 3bk^2\right] + (h) ,$$

where (h) is an infinitesimal.

Now let $\rho > 0$ be given, as small as we please, then choose $|k| < \rho$ and let h approach zero, keeping k fixed. Then since $b > 0$ it follows that $\Delta J < 0$ for all sufficiently small values of h, which proves that the line AB does not minimize the integral J.

The complete solution of the general problem which we have considered in these three chapters would require the establishment of a system of *necessary and sufficient conditions.* The above example shows that it will be necessary to add a fifth necessary condition before the complete solution of the problem is reached. We have therefore to conclude this chapter with the statement of a gap in the theory so far as it has been already developed.[1]

d) We add a *table of the various conditions* which have occurred in the problem to minimize the integral

$$J = \int_{x_0}^{x_1} F(x, y, y')\, dx \ ,$$

the end-points being fixed:

1) The minimizing curve $\mathfrak{C}_0 : y = f_0(x)$ must satisfy the differential equation

$$F_y - \frac{d}{dx} F_{y'} = 0 \ . \tag{I}$$

(*Euler's equation*, p. 22; assumptions concerning its general solution, p. 54.)

2) $$F_{y'y'}\big(x, f_0(x), f_0'(x)\big) \geqq 0 \ , \qquad \text{in } (x_0 x_1) \ . \tag{II}$$

(*Legendre's condition*, p. 47)

$$F_{y'y'}\big(x, f_0(x), \tilde{p}\big) \geqq 0 \ , \tag{IIa}$$

in $(x_0 x_1)$ for every finite $\tilde{p}$ (pp. 76 and 98).

$$F_{y'y'}(x, y, \tilde{p}) \geqq 0 \ , \tag{IIb}$$

[1] If we modify the problem by the addition of a *slope restriction, i. e.*, by subjecting the admissible curves to the further condition that their slope shall not exceed a finite fixed quantity, say

$$|y'| \leqq G \ ,$$

then the three conditions (I), (III′), (IVa′) are sufficient for a minimum.

For the function

$$\mathbf{E}_1(x, \phi(x, \gamma);\ \phi_x(x, \gamma), \tilde{p})$$

is continuous in the domain

$$\mathfrak{B}_k: \qquad x_0 \leqq x \leqq x_1 \ , \qquad |\gamma - \gamma_0| \leqq k \ , \qquad |\tilde{p}| \leqq G \ ,$$

and positive for $\gamma = \gamma_0$.

Since the domain $\mathfrak{B}_k$ is *closed*, it follows from the theorem on uniform continuity that we can take k so small that

$$\mathbf{E}_1(x, \phi(x, \gamma);\ \phi_x(x, \gamma), \tilde{p}) > 0$$

throughout the domain $\mathfrak{B}_k$, which proves the above statement.

for every (x, y) in a certain neighborhood of $\mathfrak{E}_0$ and for every finite $\tilde{p}$ (p. 96).

3)
$$x_1 \leqq x_0' \ , \qquad \text{(III)}$$

x_0' being the conjugate of x_0. (*Jacobi's condition*, pp. 58, 59, 67.)

4)
$$\mathbf{E}\left(x, f_0(x);\ f_0'(x), \tilde{p}\right) \geqq 0 \ , \qquad \text{(IV)}$$

in (x_0x_1) for every finite $\tilde{p}$. (*Weierstrass's condition*, p. 76.)

$$\mathbf{E}_1\left(x, f_0(x);\ f_0'(x), \tilde{p}\right) \geqq 0 \ , \qquad \text{(IVa)}$$

in (x_0x_1) for every finite $\tilde{p}$ (p. 76).

The omission of the equality sign in (II)–(IVa) is indicated by an accent.

Conditions (I), (II), (III) are necessary, conditions (I), (II′), (III′) are sufficient, for a weak minimum.

Conditions (I), (II), (III), (IV) are necessary, conditions (I), (IIb′), (III′) are sufficient, for a strong minimum.

§23. THE CASE OF VARIABLE END-POINTS[1]

We have so far always supposed that the two end-points

[1] Three essentially different methods have been proposed for the discussion of problems with variable end-points:

1. *The method of the Calculus of Variations proper:* It consists in computing δJ and $\delta^2 J$ either by means of Taylor's formula or by the method of differentiation with respect to ϵ, explained in §4, *b*) and *d*), and discussing the conditions $\delta J = 0$, $\delta^2 J \geqq 0$. The method was first used by LAGRANGE (1760); see *Oeuvres*, Vol. I, pp. 338, 345. He gives the general expression for δJ when the end-points are variable, viz.:

$$\delta J = \int_{x_0}^{x_1} \delta y \left(F_y - \frac{d}{dx} F_{y'}\right) dx + \Big[F\delta x + F_{y'}\delta y\Big]_0^1 \ ,$$

and derives the conditions arising from $\delta J = 0$.

The second variation for the case of variable end-points was first developed by ERDMANN (*Zeitschrift für Mathematik und Physik*, Vol. XXIII (1878), p. 364). He finds

$$\delta^2 J = \int_{x_0}^{x_1} \frac{R\,(u\delta y' - u'\delta y)^2\, dx}{u^2}$$
$$+ \left[F\delta^2 x + F_{y'}\delta^2 y + 2F_y\delta x\delta y + 2F_{y'}\delta x\delta y' + \frac{dF}{dx}\delta x^2 + \left(F_{y'y} + F_{y'y'}\frac{u'}{u}\right)\delta y^2\right]_0^1 ,$$

where u is an integral of Jacobi's differential equation. By considering such spe-

of the required curve are fixed. In this section we propose to consider the modification of the problem in which one of the end-points, say 0, is fixed, whilst the other, 1, is movable on a given curve $\tilde{\mathfrak{C}}$.

Suppose the curve

$$\mathfrak{C}_0: \qquad y = f_0(x) \;, \qquad x_0 \leqq x \leqq x_1 \;,$$

—which we suppose to be of class C' and to lie in the interior of the region $\mathfrak{R}$—minimizes the integral J with these

cial variations for which $\delta y = Cu$, he makes the integral vanish and thus reduces the question to the discussion of the sign of the remaining function of the variations δx_i, δy_i, $\delta^2 x_i$, $\delta^2 y_i$. These variations are connected by relations which depend upon the special nature of the initial conditions. For instance, for the initial conditions considered in the text the expression for $\delta^2 J$ reduces to the expression (36) for $J''(x_1)$ multiplied by δx_1^2.

For the general integral

$$J = \int_{x_0}^{x_1} F(x, y_1, y_2, \cdots, y_n, y_1', y_2', \cdots, y_n')\, dx \;,$$

where $y_1, y_2, \cdots y_n$ are connected by a number of finite or differential relations, the second variation in the case of variable end-points was studied by A. MAYER, *Leipziger Berichte* (1896), p. 436; for the integral in parameter-representation

$$J = \int_{t_0}^{t_1} F(x, y, x', y')\, dt$$

by BLISS, *Transactions of the American Mathematical Society*, Vol. III (1902), p. 132 (compare §30).

2. *The method of Differential Calculus:* This method is explained in a general way in DIENGER'S *Grundriss der Variationsrechnung* (1867). It decomposes the problem into two problems by first considering variations which leave the end-points fixed, and then variations which vary the end-points, the neighboring curves considered being themselves extremals. The second part of the problem reduces to a problem of the theory of ordinary maxima and minima. This method has been used by A. MAYER in an earlier paper on the second variation in the case of variable end-points for the general type of integrals mentioned above (*Leipziger Berichte* (1884), p. 99). It is superior to the first method not only on account of its greater simplicity and its more elementary character, but because—by utilizing the well-known sufficient conditions for ordinary maxima and minima—it leads, in a certain sense, to sufficient conditions if combined with WEIERSTRASS'S sufficient conditions for the case of fixed end-points. For these reasons I have adopted this method in the text.

3. *Kneser's method:* This method, which has been developed by KNESER in his *Lehrbuch*, is based upon an extension of certain well-known theorems on geodesics. It leads in the simplest way to sufficient conditions, but must be supplemented by one of the two preceding methods for an exhaustive treatment of the necessary conditions. A detailed account of this method will be given in Chapter V.

initial conditions. Then we must have $\Delta J \geqq 0$ for every curve $\overline{\mathfrak{C}}$ of class D′ which begins at the point 0 and ends at a point of the curve $\tilde{\mathfrak{C}}$ and which lies moreover in a certain

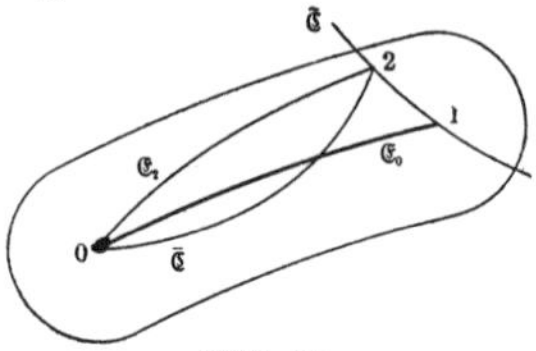

FIG. 21

neighborhood[1] $\mathfrak{U}$ of $\mathfrak{E}_0$.

a) Among the totality of these "admissible curves" we consider in the first place those which end at the point 1. For these also the inequality $\Delta J \geqq 0$ must hold, and therefore all the conditions which we have found to be necessary in the case of fixed end-points must be fulfilled in the present case.

The arc $\mathfrak{E}_0$ *must therefore be an extremal, Legendre's condition*

$$F_{y'y'} \geqq 0 \tag{II}$$

must be satisfied along $\mathfrak{E}_0$, *and the conjugate point* 0′ *to* 0 *must not lie between* 0 *and* 1.

We suppose in the sequel that the arc $\mathfrak{E}_0$ is an extremal, that the condition

$$F_{y'y'}(x, y, \tilde{p}) > 0 \tag{IIb′}$$

is fulfilled at every point (x, y) of a certain neighborhood of $\mathfrak{E}_0$ for every finite value of $\tilde{p}$ and that the arc $\mathfrak{E}_0$ does not contain the conjugate point 0′ (Condition III′).

b) Further necessary conditions are obtained by considering variations which do vary the end-point 1. Various methods[2] have been proposed for this purpose. The following elementary method reduces the further discussion to a problem of *ordinary maxima and minima:*

If the extremal $\mathfrak{E}_0$ minimizes the integral J in the sense explained above, then $\mathfrak{E}_0$ must, in particular, furnish a smaller

[1] Compare §3, *b*); we may for instance choose for $\mathfrak{U}$ the special neighborhood (ρ) used in the problem with fixed end-points (§3, *c*)), increased by a semi-circle of radius ρ with the point 1 for center.

[2] Compare footnote 1, p. 102.

value than (or at most the same value as) every extremal which can be drawn from the point 0 to the curve $\tilde{\mathfrak{C}}$ and which lies in a certain neighborhood of $\mathfrak{E}_0$.

And since under the above assumptions (IIb′) and (III′) each of these extremals—(when its end-points are considered as fixed)—minimizes the integral J, it seems[1] self-evident that also the converse is true.

Let then

$$y = \phi(x, \gamma) \tag{30}$$

represent the set of extremals through the point 0, and let γ_0 denote again the value of γ which corresponds to $\mathfrak{E}_0$. From the above assumptions (IIb′) and (III′) it follows that this set furnishes for $|\gamma - \gamma_0| \leqq k$ an (improper) field[2] $\mathfrak{F}_k$ about the arc $\mathfrak{E}_0$ if k is taken sufficiently small.

Hence, if $2: (x_2, y_2)$ be any point of the curve $\tilde{\mathfrak{C}}$ in a certain vicinity of the point 1, then there passes one and but one extremal

$$\mathfrak{E}_2: \qquad y = \phi(x, \gamma_2)$$

of the field through the point 2. The parameter γ_2 is a single-valued function of x_2, y_2 of class C': $\gamma_2 = \psi(x_2, y_2)$.

If

$$\tilde{y} = \tilde{f}(x)$$

is the equation of the given curve, which we suppose to be of class C'', then $y_2 = \tilde{f}(x_2)$ and $\gamma_2 = \psi(x_2, \tilde{f}(x_2))$.

Hence the integral J taken along the extremal $\mathfrak{E}_2$ from the point 0 to the point 2 is a single-valued function of x_2, say

[1] It will be seen under *e*) how far this conclusion is correct.

[2] Compare p. 83, footnote 2. In the present case the field $\mathfrak{F}_k$ consists of all points (x, y) furnished by (30), when x, γ are restricted to the domain

$$x_0 \leqq x \leqq X_1, \quad |\gamma - \gamma_0| \leqq k,$$

where X_1 is some value greater than x_1; k is supposed to be taken so small that (IIb′) holds throughout $\mathfrak{F}_k$ and that $\phi_\gamma(x, \gamma) \neq 0$ throughout the domain

$$x_0 < x \leqq X_1, \quad |\gamma - \gamma_0| \leqq k.$$

$$J(x_2)=\int_{x_0}^{x_2} F\big(x,\, \phi(x,\gamma_2),\, \phi_x(x,\gamma_2)\big)\,dx\ .$$

And this function $J(x_2)$ must have a minimum for $x_2=x_1$. Therefore we must have

$$J'(x_1)=0\ ,\qquad J''(x_1)\geqq 0\ . \tag{31}$$

c) The derivative of the integral $J(x_2)$ has already been computed[1] in §20, *c*) (equation (24)). Accordingly

$$J'(x_2)=F(x_2,\, y_2,\, p_2)+(\tilde{p}_2-p_2)\,F_{y'}(x_2,\, y_2,\, p_2)\ , \tag{32}$$

where $p_2=\phi_x(x_2,\gamma_2)$ is the slope of the extremal $\mathfrak{E}_2$, and $\tilde{p}_2=\tilde{f}'(x_2)$ the slope of the curve $\tilde{\mathfrak{C}}$, at the point 2.

Hence we obtain the result:

The co-ordinates x, y of the movable end-point must satisfy the condition[2]

$$F(x_1,\, y_1,\, y_1')+(\tilde{y}_1'-y_1')\,F_{y'}(x_1,\, y_1,\, y_1')=0\ , \tag{33}$$

where y_1' and $\tilde{y}_1'$ refer to the extremal $\mathfrak{E}_0$ and to the curve $\mathfrak{C}$ respectively.

If this condition is satisfied we shall say that *the curve $\mathfrak{C}$ is* TRANSVERSE[3] *to the extremal $\mathfrak{E}_0$ at the point 1.*

Equation (33) together with the two equations

$$f(x_0,\alpha,\beta)=y_0\ ,\qquad f(x_1,\alpha,\beta)=\tilde{f}(x_1)\ ,$$

determine in general the two constants of integration α, β in the general solution of Euler's differential equation, as well as the abscissa x_1 of the point 1.

We suppose in the sequel that condition (33) is fulfilled.

d) We next proceed to the computation of $J''(x_2)$. From (32) we obtain

[1] We suppose that the co-ordinates of the movable end-point do not occur explicitly in the function $F(x,y,y')$; if they do occur, another term must be added to the expression of $J'(x_2)$. Compare for this case KNESER, *Lehrbuch*, §12. An example of this exceptional case is the brachistochrone; compare LINDELÖF-MOIGNO, *Calcul des variations*, No. 113, and the references given in PASCAL, *Variationsrechnung*, §31.

[2] In accordance with §8, end.

[3] In the use of the word "transverse" I follow OSGOOD, *loc. cit.*, p. 112. KNESER, who first introduced the term (*Lehrbuch*, §10), used it with a slightly different meaning; he says: the extremal $\mathfrak{E}_0$ is transverse to the curve $\tilde{\mathfrak{C}}$ if (33) is satisfied.

$$J''(x_2) = F_x + F_y \frac{dy_2}{dx_2} + F_{y'} \frac{dp_2}{dx_2} + F_{y'} \left(\frac{d\tilde{p}_2}{dx_2} - \frac{dp_2}{dx_2} \right)$$
$$+ (\tilde{p}_2 - p_2) \left[F_{y'x} + F_{y'y} \frac{dy_2}{dx_2} + F_{y'y'} \frac{dp_2}{dx_2} \right] .$$

But

$$y_2 = \tilde{f}(x_2) = \phi(x_2, \gamma_2) ,$$
$$p_2 = \phi_x(x_2, \gamma_2) , \qquad \tilde{p}_2 = \tilde{f}'(x_2) ;$$

hence

$$\frac{dy_2}{dx_2} = \tilde{f}'(x_2) , \qquad \frac{d\tilde{p}_2}{dx_2} = \tilde{f}''(x_2) , \tag{34}$$
$$\frac{dp_2}{dx_2} = \phi_{xx}(x_2, \gamma_2) + \phi_{x\gamma}(x_2, \gamma_2) \frac{d\gamma_2}{dx_2} ,$$

and $\frac{d\gamma_2}{dx_2}$ is determined by

$$\tilde{f}'(x_2) = \phi_x(x_2, \gamma_2) + \phi_\gamma(x_2, \gamma_2) \frac{d\gamma_2}{dx_2} .$$

Substituting these values for

$$\frac{dy_2}{dx_2} , \qquad \frac{dp_2}{dx_2} , \qquad \frac{d\tilde{p}_2}{dx_2} ,$$

and remembering that on account of Euler's equation

$$F_{y'x} = F_y - F_{y'y} \phi_x - F_{y'y'} \phi_{xx} ,$$

we obtain for $x = x_1$ the following result:[1]

Let A_1 and B_1 denote the expressions:

$$A_1 = F_x + (2\tilde{y}_1 - y_1') F_y + \tilde{y}_1'' F_{y'} + (\tilde{y}_1' - y_1')^2 F_{y'y} , \tag{35}$$
$$B_1 = (\tilde{y}_1' - y_1')^2 F_{y'y'} ,$$

the arguments of the derivatives of F being x_1, y_1, y_1'; then

$$J''(x_1) = A_1 + B_1 \frac{\phi_{\gamma x}(x_1, \gamma_0)}{\phi_\gamma(x_1, \gamma_0)} . \tag{36}$$

For the further discussion of the inequality $J''(x_1) \geqq 0$, we leave aside the exceptional case where $\tilde{y}_1' = y_1'$, *i. e., we suppose that the extremal* $\mathfrak{E}_0$ *and the curve* $\tilde{\mathfrak{C}}$ *are not tangent to each other at the point 1.* Then $B_1 > 0$, since we have moreover already supposed that $F_{y'y'} > 0$.

[1] Given, in a slightly different form, by BLISS, *Mathematische Annalen*, Vol. LVIII (1903), p. 77.

According to equation (30) of §15, we have in the notation of §§13 and 14:

$$\phi_\gamma(x_1, \gamma_0) = C\Delta(x_1, x_0)$$

and therefore

$$\phi_{\gamma x}(x_1, \gamma_0) = C\frac{\partial\Delta(x_1, x_0)}{\partial x_1}.$$

Now let $\mathrm{H}(x_1, x)$ denote the function

$$\mathrm{H}(x_1, x) = A_1\Delta(x_1, x) + B_1\frac{\partial\Delta(x_1, x)}{\partial x_1}, \tag{37}$$

then the expression for $J''(x_1)$ may be written

$$J''(x_1) = \frac{\mathrm{H}(x_1, x_0)}{\Delta(x_1, x_0)}.$$

The function

$$\Delta(x_1, x) = r_1(x_1)\, r_2(x) - r_2(x_1)\, r_1(x)$$

is an integral of Jacobi's differential equation and vanishes for $x = x_1$. The function $\mathrm{H}(x_1, x)$ is likewise an integral of Jacobi's differential equation, since it is linearly expressible in terms of $r_1(x)$ and $r_2(x)$. Since $B_1 > 0$ and

$$r_1(x_1)\, r_2'(x_1) - r_2(x_1)\, r_1'(x_1) \neq 0 \tag{38}$$

(see pp. 57, 58),

$$\mathrm{H}(x_1, x_1) \neq 0. \tag{39}$$

Hence if we denote by x_1' the root of the equation

$$\Delta(x_1, x) = 0$$

next smaller than x_1 and by x_1'' the root next smaller than x_1, of the equation

$$\mathrm{H}(x_1, x) = 0\text{[1]},$$

it follows from Sturm's theorem[2] that

$$x_1'' > x_1'.$$

At $x = x_1''$, $\mathrm{H}(x_1, x)$ changes sign.

[1] Compare p. 58.

[2] Compare p. 58, footnote 2. This remark is due to BLISS, *Transactions*, etc. p. 138.

Again from (38) it follows that

$$\underset{x=x_1}{L}(x_1 - x)\frac{\partial \Delta(x_1, x)}{\partial x_1}\Big/\Delta(x_1, x) = 1$$

and therefore

$$\underset{x=x_1-0}{L}\left(A_1 + B_1\frac{\partial \Delta(x_1, x)}{\partial x_1}\Big/\Delta(x_1, x)\right) = +\infty\ .$$

Hence we infer that

$$J''(x_1)\begin{cases} >0 & \text{when } x_1'' < x_0 < x_1\ , \\ =0 & \text{when } x_0 = x_1''\ , \\ <0 & \text{when } x_1' < x_0 < x_1''\ . \end{cases}$$

For reasons which will appear later on (under *f*), the point of the extremal $\mathfrak{E}_0$ whose abscissa is x_1'' is called, according to KNESER,[1] the "*focal point*" of the curve $\tilde{\mathfrak{C}}$ on the extremal $\mathfrak{E}_0$.

We have therefore reached the theorem: *For a minimum it is necessary that the focal point of the curve $\tilde{\mathfrak{C}}$ on the extremal $\mathfrak{E}_0$ shall not lie between the points 0 and 1.*

e) It remains to consider the question of the *sufficiency* of these conditions.

If in addition to (IIb′) and (33) the condition

$$x_1'' < x_0 \tag{41}$$

is satisfied, then

$$J'(x_1) = 0\ , \qquad J''(x_1) > 0\ ,$$

and therefore the function $J(x_2)$ has a minimum for $x_2 = x_1$.

Let now $\bar{\mathfrak{C}}$ be any curve of class D' which begins at the point 0 and ends at some point 2 of $\tilde{\mathfrak{C}}$, and which lies moreover in the improper field $\mathfrak{F}_k$ about $\mathfrak{E}_0$ defined under *b*). Let $\mathfrak{E}_2$ be the extremal of the field from the point 0 to the point 2 (see Fig. 21), then we have

$$J_{\mathfrak{E}_0} < J_{\mathfrak{E}_2}\ .$$

[1] The discovery of the focal point ("Brennpunkt") is due to KNESER, see *Lehrbuch*, §24. For the special case of the straight line, the focal point occurs already in ERDMANN's paper referred to above. BLISS uses "critical point" for "Brennpunkt."

On the other hand, since we have supposed k so small that $\phi_\gamma(x, \gamma) \neq 0$ for

$$x_0 < x \leqq X_1 , \qquad |\gamma - \gamma_0| \leqq k ,$$

the region $\mathfrak{S}_k$ is at the same time an (improper) field[1] about the extremal $\mathfrak{E}_2$ and therefore since (IIb′) holds throughout $\mathfrak{S}_k$,

$$J_{\mathfrak{E}_2} < J_{\bar{\mathfrak{C}}} ,$$

according to §22, *b*). Hence

$$J_{\mathfrak{E}_0} < J_{\bar{\mathfrak{C}}} .$$

The extremal $\mathfrak{E}_0$ furnishes therefore a smaller value for the integral J than any other curve of class D' which can be drawn in the region $\mathfrak{S}_k$ from the point 0 to the curve $\tilde{\mathfrak{C}}$, and *in this sense the extremal $\mathfrak{E}_0$ minimizes*[2] *the integral J if the conditions (IIb′), (33) and (41) are fulfilled.*

EXAMPLE VIa: *To draw the curve of shortest length from a given point to a given curve.*

Here:

$$F = \sqrt{1 + y'^2} ;$$

hence we obtain for the condition of transversality

$$1 + y_1' \tilde{y}_1' = 0 ,$$

i. e., the minimizing straight line must be *normal* to the curve $\tilde{\mathfrak{C}}$ at the point 1.

Further we get easily

$$\mathrm{H}(x_1, x) = \frac{y_1' \tilde{y}_1''}{\sqrt{1 + y_1'^2}}(x_1 - x) + \frac{(\tilde{y}_1' - y_1')^2}{(\sqrt{1 + y_1'^2})^3} ;$$

therefore

$$x_1'' = x_1 - \frac{\tilde{y}'(1 + \tilde{y}'^2)}{\tilde{y}_1''} .$$

[1] In the discussion concerning the construction of a field about $\mathfrak{E}_0$ in §19, we have for simplicity restricted γ to an interval $(\gamma_0 - k, \gamma_0 + k)$ whose middle point is $\gamma = \gamma_0$. We might just as well have taken an interval of the more general form $(\gamma_0 - k_1, \gamma_0 + k_2)$. In the present case the term field must be understood in this slightly more general sense.

[2] It should, however, be observed that the region $\mathfrak{S}_k$ does not, strictly speaking, constitute a neighborhood (see §3, *b*)) of the arc $\mathfrak{E}_0$ since its width approaches zero as x approaches the value x_0. The proof that $\mathfrak{E}_0$ minimizes the integral J is therefore not quite complete. KNESER's sufficiency proof, which will be given in chap. V for the problem in parameter-representation, is not open to this objection.

Hence it follows that *the center of curvature 1'' of the curve $\tilde{\mathfrak{C}}$ at the point 1 must not lie between the point 0 and the point 1.* Conversely: If this condition is fulfilled and if moreover 1'' does not coincide with the point 0, then the straight line 01 actually furnishes a minimum.

Entirely analogous results are obtained in the case when the point 1 is fixed and the point 0 movable on a given curve. The condition of transversality must be satisfied at the point 0. Again, if A_0, B_0 have the same meaning for the point 0 as the constants A_1, B_1 for the point 1, and if x_0'' denotes the root next[1] greater than x_0 of the equation

$$\mathrm{H}(x_0, x) \equiv A_0 \Delta(x_0, x) + B_0 \frac{\partial \Delta(x_0, x)}{\partial x_0} = 0 , \tag{42}$$

then x_0'' must not be less than x_1.

f) Geometrical interpretation of the focal point. Let us consider the problem to construct through a point 2 of the curve $\tilde{\mathfrak{C}}$ in the vicinity of the point 1 an extremal which shall be cut transversely at the point 2 by the curve $\tilde{\mathfrak{C}}$. Let

$$y = f(x, \alpha, \beta)$$

be the required extremal. Then we have for the determination of α and β the two equations

$$\begin{aligned} M &\equiv f(x_2, \alpha, \beta) - \tilde{f}(x_2) = 0 , \\ N &\equiv F(x_2, y_2, \tilde{q}_2) + (\tilde{p}_2 - \tilde{q}_2) F_{y'}(x_2, y_2, \tilde{q}_2) = 0 , \end{aligned} \tag{43}$$

where

$$y_2 = \tilde{f}(x_2) , \qquad \tilde{p}_2 = \tilde{f}'(x_2) , \qquad \tilde{q}_2 = f_x(x_2, \alpha, \beta) .$$

The two equations (43) are satisfied for $x_2 = x_1$, $\alpha = \alpha_0$, $\beta = \beta_0$, since $\tilde{\mathfrak{C}}$ is transverse to $\mathfrak{C}_0$ at the point 1; the left-hand sides of the two equations (43) are functions of x_2, α, β of class C' in the vicinity of $x_2 = x_1$, $\alpha = \alpha_0$, $\beta = \beta_0$ and their Jacobian with respect to α and β is different from zero for $x_2 = x_1$, $\alpha = \alpha_0$, $\beta = \beta_0$, if $\tilde{y}_1' - y_1' \neq 0$ as we have supposed; for it reduces to

$$(\tilde{y}' - y_1') F_{y'y'} \left(r_1(x_1) r_2'(x_1) - r_2(x_1) r_1'(x_1) \right) .$$

[1] Compare p. 58.

Hence the equations (43) admit, according to the theorem on implicit functions,[1] a unique solution:

$$\alpha = \alpha(x_2) \ , \qquad \beta = \beta(x_2) \ ,$$

which is of class C' in the vicinity of $x_2 = x_1$ and satisfies the initial conditions

$$\alpha(x_1) = \alpha_0 \ , \qquad \beta(x_1) = \beta_0 \ .$$

If we denote

$$f\big(x, \ \alpha(x_2), \ \beta(x_2)\big) = g(x, x_2) \ ,$$

the required extremal is therefore

$$y = g(x, x_2) \ , \tag{44}$$

and if we consider x_2 as a variable parameter, this equation represents a set of extremals each of which is cut transversely by the curve $\tilde{\mathfrak{C}}$; the extremal $\mathfrak{C}_0$ is itself contained in the set and corresponds to $x_2 = x_1$.

The envelope $\mathfrak{F}$ of the set (44) is defined by the two equations

$$y = g(x, x_2) \ , \qquad g_{x_2}(x, x_2) = 0 \ ,$$

and the abscissae of the points at which the extremal $\mathfrak{C}_0$ meets this envelope are the roots of the equation

$$g_{x_2}(x, x_2)\Big|^{x_2=x_1} = 0 \ .$$

To obtain this equation we compute the derivatives

$$\frac{d\alpha}{dx_2} \ , \qquad \frac{d\beta}{dx_2}$$

from the two equations $dM/dx_2 = 0$, $dN/dx_2 = 0$, substitute their values in the equation

$$g_{x_2}(x, x_2) \equiv f_\alpha \frac{d\alpha}{dx_2} + f_\beta \frac{d\beta}{dx_2} = 0 \ ,$$

and finally put $x_2 = x_1$, $\alpha = \alpha_0$, $\beta = \beta_0$.

Carrying out this process, we are led to the three equations

[1] Compare footnote 2, p. 35.

$$r_1(x)\,\alpha'(x_1)+r_2(x)\,\beta'(x_1) \;=0\;,$$
$$r_1(x_1)\,\alpha'(x_1)+r_2(x_1)\,\beta'(x_1) \;=\tilde{y}_1'-y_1'\;,$$
$$(\tilde{y}_1'-y_1')\,F_{y'y'}\Big[r_1'(x_1)\,\alpha'(x_1)+r_2'(x_1)\,\beta'(x_1)\Big]=-A_1\;,$$

from which, by eliminating $\alpha'(x_1)$, $\beta'(x_1)$, we obtain the result

$$\mathrm{H}(x_1, x)=0\;, \qquad i.\,e.,$$

The focal point[1] *is the point at which the extremal* $\mathfrak{E}_0$ *touches for the first time—counting from the point 1 toward the point 0—the envelope of the set of extremals which are cut transversely by the curve* $\tilde{\mathfrak{C}}$.

EXAMPLE VIa: The set (44) consists of the normals to the curve $\tilde{\mathfrak{C}}$; the envelope $\mathfrak{F}$ is the evolute of the curve $\tilde{\mathfrak{C}}$.

g) Case of two movable end-points: We add a few remarks concerning the case when the point 0 is movable on a curve $\tilde{\mathfrak{C}}_0$ and at the same time the point 1 movable on a curve $\tilde{\mathfrak{C}}_1$.

The consideration of special variations leads at once to the result that the minimizing curve must be an extremal, that the condition of transversality must hold at both end-points, and that the inequalities

$$x_0 \geqq x_1''\,, \qquad x_1 \leqq x_0''$$

must be satisfied.

But still another condition must be added: If x_1''' denotes the root next greater than x_1 of the equation

$$\mathrm{H}(x_1, x)=0\;,$$

then *the following inequality must be satisfied:*[2]

[1] This geometrical interpretation of the focal point is due to KNESER; see *Lehrbuch*, §24.

[2] This result is due to BLISS; see *Mathematische Annalen*, Vol. LVIII (1903), p. 70. He also proves that for a regular problem the condition $x_1 < x_1''' < x_0''$, together with the two transversality conditions and the condition that the minimizing curve is an extremal, are sufficient for a minimum. His proof is based upon Kneser's theory of the problem with one variable end-point.

For the example of the curve of shortest length between two given curves, the inequality (45) had already been given by ERDMANN (*loc. cit.*). Another important example with both end-points variable (the special isoperimetric problem) has been completely discussed by KNESER (*Mathematische Annalen*, Vol. LVI (1902), p. 169).

$$x_1 < x_1''' \leqq x_0'' . \tag{45}$$

The problems on variable end-points which we have discussed in this section are special cases of the problem: To minimize the integral J when the co-ordinates of the two end-points are connected by a number of relations:[1]

$$\Phi_\nu(x_0, y_0, x_1, y_1) = 0 .$$

The "method of differential calculus" used in this section can be applied also to this case.

The number of independent relations cannot exceed four; if it is exactly equal to four, we have the case of fixed end-points. If both end-points are perfectly unrestricted, the vanishing of the first variation leads to the four conditions

$$F\Big|^1 = 0 , \quad F\Big|^0 = 0 , \quad F_{y'}\Big|^1 = 0 , \quad F_{y'}\Big|^0 = 0 ,$$

which are in general incompatible.

[1] Compare KNESER, *Lehrbuch*, §10.

CHAPTER IV

WEIERSTRASS'S THEORY OF THE PROBLEM IN PARAMETER-REPRESENTATION[1]

§24. FORMULATION OF THE PROBLEM

IN the previous chapters we have confined ourselves to curves which are representable in the form $y=f(x)$, a restriction of a very artificial character in all truly geometrical problems. We are now going to remove this restriction by assuming henceforth all curves expressed in parameter-representation.

a) Generalities concerning curves in parameter-representation.[2]

A "*continuous curve*" $\mathfrak{C}$ is defined by a system of two equations

$$\mathfrak{C}: \qquad x=\phi(t), \qquad y=\psi(t), \qquad t_0 \leqq t \leqq t_1, \qquad (1)$$

ϕ and ψ being functions of t, defined and continuous in $(t_0 t_1)$. As t increases from t_0 to t_1, the curve is described in

[1] The treatment of the problems of the Calculus of Variations in parameter-representation is entirely due to WEIERSTRASS; he used it in his lectures at least as early as 1872. In order to avoid repetitions, we shall discuss in detail only those points in which the new treatment differs essentially from the old one. For the rest, we shall confine ourselves to an account of the results.

As regards the *relative merits of the two methods*, one is inclined to consider the older method—in which x is taken for the independent variable—as antiquated and imperfect when compared with Weierstrass's method; unjustly, however, for the two methods deal with two clearly distinct problems, and which of the two deserves the preference, depends upon the nature of the special problem under consideration.

Generally speaking one may say that in all *truly geometrical problems* the method of parameter-representation is not only preferable, but is the only one which furnishes a complete solution. On the other hand, the older method has to be applied whenever a *function* of minimizing properties is to be determined (for instance, *Dirichlet's problem*).

For examples illustrating the relation between the two methods, see BOLZA, *Bulletin of the American Mathematical Society* (2), Vol. IX (1903), p. 6.

[2] Compare J. I, Nos. 96–113.

a certain sense, called the "positive sense," from its origin, say 0, to its end-point, say 1.

If we make the "*parameter-transformation*":

$$t = \chi(\tau) \;, \tag{2}$$

where $\chi(\tau)$ is a continuous function of τ which constantly increases from t_0 to t_1 as τ increases from τ_0 to τ_1, the equations (1) are changed into

$$x = \phi\left(\chi(\tau)\right) = \Phi(\tau) \;, \qquad y = \psi\left(\chi(\tau)\right) = \Psi(\tau) \;. \tag{1a}$$

Vice versa, the equations (1a) are again transformed into (1) by the inverse transformation

$$\tau = \chi^{-1}(t) \;. \tag{2a}$$

We agree to consider the two curves defined by (1) and (1a) as identical, and conversely two curves will be considered as identical only[1] when their equations can be transformed into each other by a parameter-transformation of the above properties.

The curve $\mathfrak{C}$ will be said to be *of class* $C'(C'')$ if the parameter t can be so selected that $\phi(t)$ and $\psi(t)$ have *continuous first* (*and second*) *derivatives in* (t_0t_1), and if moreover ϕ' and ψ' do not vanish simultaneously in (t_0t_1) so that

$$\phi'^2 + \psi'^2 \neq 0 \qquad \text{in } (t_0t_1) \;. \tag{3}$$

A curve of class C' has at every point a *continuously turning tangent;* the amplitude θ of its positive direction is given by the equations

$$\cos\theta = \frac{\phi'}{\sqrt{\phi'^2 + \psi'^2}} \;, \qquad \sin\theta = \frac{\psi'}{\sqrt{\phi'^2 + \psi'^2}} \;. \tag{4}$$

Every curve of class C' is rectifiable,[2] and the length s of the arc t_0t is expressible by the definite integral

[1] According to this agreement, a curve (more exactly "path-curve," E. H. MOORE) is not simply the totality of points defined by (1) but the totality of these points *taken in the order defined by* (*1*).

[2] Compare J. I, Nos. 105–111.

$$s=\int_{t_0}^{t}\sqrt{\phi'^2+\psi'^2}\,dt\ . \tag{5}$$

By an "*ordinary curve*" will be understood a continuous curve which is either of class C' or else made up of a finite number of arcs of class C'. A point where two different arcs meet will be called a "corner" if the direction of the positive tangent undergoes a discontinuity at that point. A curve will be said to be *regular* at a point $t=t'$, if for sufficiently small values of $|t-t'|$, x and y are expansible into convergent power-series:

$$x=a+a_1(t-t')+a_2(t-t')^2+\cdots\ ,$$
$$y=b+b_1(t-t')+b_2(t-t')^2+\cdots\ ,$$

and if moreover a_1 and b_1 are not both zero.

b) Integrals taken along a curve; conditions for their invariance under a parameter-transformation.

Let $F(x, y, x', y')$ be a function of four independent variables which is of class C''' in a domain $\mathfrak{T}$ which consists of all points x, y, x', y' for which *a*) x, y lies in a certain region $\mathfrak{R}$ of the x, y-plane, *b*) x', y' are not both zero.

We suppose that the curve $\mathfrak{C}$ defined by (1) lies entirely in $\mathfrak{R}$, and select two points 2 and 3 ($t_2 < t_3$) on $\mathfrak{C}$. Then we consider the definite integral

$$J=\int_{t_2}^{t_3}F(x, y, x', y')\,dt\ ,$$

in which x, y, x', y' are replaced by $\phi(t)$, $\psi(t)$, $\phi'(t)$, $\psi'(t)$ respectively, and ask: *Under what conditions will the value of the integral J depend only on the arc 23 and not on the choice of the parameter t?*

The simplest example of an integral which is independent of the choice of the parameter is the length of the arc 23, which is always expressed by the definite integral

$$\int_{t_2}^{t_3}\sqrt{x'^2+y'^2}\,dt\ ,$$

no matter what quantity has been selected for the independent variable t, provided that $t_2 < t_3$, so that if we pass from the parameter t to another parameter τ by any admissible transformation (2), we must have

$$\int_{t_2}^{t_3} \sqrt{\left(\frac{dx}{dt}\right)^2 + \left(\frac{dy}{dt}\right)^2}\, dt = \int_{\tau_2}^{\tau_3} \sqrt{\left(\frac{dx}{d\tau}\right)^2 + \left(\frac{dy}{d\tau}\right)^2}\, d\tau \,.$$

Returning now to the general case, our question may be formulated explicitly as follows:

Under what conditions is

$$\int_{t_2}^{t_3} F\left(x, y, \frac{dx}{dt}, \frac{dy}{dt}\right) dt = \int_{\tau_2}^{\tau_3} F\left(x, y, \frac{dx}{d\tau}, \frac{dy}{d\tau}\right) d\tau \,, \tag{6}$$

with the understanding that this relation is to hold:

a) For every transformation $t = \chi(\tau)$ of the properties indicated above;

β) For all positions of the two points 2 and 3 on the curve $\mathfrak{C}$;

γ) For all possible curves $\mathfrak{C}$ of class C', lying in $\mathfrak{R}$?

On account of β) we may differentiate (6) with respect to τ_3; writing for brevity t, τ instead of t_3, τ_3, we obtain

$$F\left(x, y, \frac{dx}{dt}, \frac{dy}{dt}\right) \frac{dt}{d\tau} = F\left(x, y, \frac{dx}{d\tau}, \frac{dy}{d\tau}\right) ,$$

or since

$$\frac{dx}{d\tau} = \frac{dx}{dt}\frac{dt}{d\tau} \,, \qquad \frac{dy}{d\tau} = \frac{dy}{dt}\frac{dt}{d\tau} \,:$$

$$F\left(x, y, \frac{dx}{dt}, \frac{dy}{dt}\right) \frac{dt}{d\tau} = F\left(x, y, \frac{dx}{dt}\frac{dt}{d\tau}, \frac{dy}{dt}\frac{dt}{d\tau}\right) . \tag{7}$$

On account of *a*) this must hold for the special transformation

$$t = k\tau \,,$$

k being a positive constant. Hence

$$F\left(x, y, k\frac{dx}{dt}, k\frac{dy}{dt}\right) = k\,F\left(x, y, \frac{dx}{dt}, \frac{dy}{dt}\right) .$$

But by properly choosing the curve (1) (see assumption γ)) and the parameter t, we can give the four quantities

$$x,\, y,\, \frac{dx}{dt},\, \frac{dy}{dt}$$

any arbitrary system of values in the domain $\mathfrak{T}$, and therefore the relation

$$F(x, y, kx', ky') = kF(x, y, x', y') \tag{8}$$

must hold identically for all values of the independent variables x, y, x', y' in $\mathfrak{T}$ and for all positive values of k, or as we shall say: *$F(x, y, x', y')$ must be "positively homogeneous" and of dimension one with respect to x', y'.*

Vice versa, if this condition is satisfied, (7) holds since we suppose

$$\frac{dt}{d\tau} > 0 ,$$

and therefore also (6), as follows by integrating (7) between the limits τ_2 and τ_3. This shows that *the homogeneity condition (8) is necessary and sufficient for the invariance of the integral J.*[1]

We shall in the sequel always suppose that the function F satisfies the homogeneity condition (8), and we shall denote the value of the integral

$$\int_{t_0}^{t_1} F\big(\phi(t), \psi(t), \phi'(t), \psi'(t)\big)\, dt$$

indifferently by $J_{\mathfrak{C}}$ or J_{01}, and call it the integral of the function $F(x, y, x', y')$ taken along the curve $\mathfrak{C}$.

If we wish to reverse[2] the direction of integration we must first introduce a new parameter which increases as the

[1] WEIERSTRASS, *Lectures;* also KNESER, *Lehrbuch*, §3.

This lemma has been extended to the case where F contains higher derivatives of x and y by ZERMELO, *Dissertation*, pp. 2–23; to the case of double integrals by KOBB, *Acta Mathematica*, Vol. XVI (1892), p. 67.

[2] Compare KNESER, *Lehrbuch*, p. 9.

curve is described from the point 1 to the point 0, for instance: $u=-t$. The equations

$$\mathfrak{C}^{-1}: \qquad x=\phi(-u), \quad y=\psi(-u), \quad u_0 \leqq u \leqq u_1,$$

where $u_0=-t_1$, $u_1=-t_0$, represent the same totality of points as (1), but the sense is reversed.

The integral of $F(x, y, x', y')$ taken along $\mathfrak{C}^{-1}$ has the value

$$J_{10}=\int_{u_0}^{u_1} F\left(x, y, \frac{dx}{du}, \frac{dy}{du}\right) du,$$

$$=\int_{u_0}^{u_1} F\big(\phi(-u), \psi(-u), -\phi'(-u), -\psi'(-u)\big)\, du,$$

$$=\int_{t_0}^{t_1} F\big(\phi(t), \psi(t), -\phi'(t), -\psi'(t)\big)\, dt.$$

If the relation (8) holds also for negative values of k, as happens, for instance, when F is a rational function of x', y', then

$$F(x, y, -x', -y') = -F(x, y, x', y'),$$

and therefore: $J_{10}=-J_{01}$.

But the relation (8) need not hold for negative values of k; thus in the example of the length we have for negative values of k

$$F(x, y, kx', ky') = -kF(x, y, x', y');$$

hence in this case $J_{10}=J_{01}$.

In other cases the relation is more complicated, for instance, when

$$F = xy' - x'y + \lambda\sqrt{x'^2+y'^2}.$$

From the homogeneity condition (8) follow a number of important *relations between the partial derivatives of* F.

Differentiating (8) with respect to k and then putting $k=1$, we get

$$x'F_{x'} + y'F_{y'} = F. \tag{9}$$

Differentiating this relation with respect to x and y, we obtain

$$F_x = x'F_{x'x} + y'F_{y'x}, \qquad F_y = x'F_{x'y} + y'F_{y'y}. \tag{10}$$

Differentiating (9) with respect to x' and y' we get

$$x'F_{x'x'} + y'F_{y'x'} = 0 \ , \qquad x'F_{x'y'} + y'F_{y'y'} = 0 \ ;$$

hence if x' and y' are not both zero,

$$F_{x'x'} : F_{x'y'} : F_{y'y'} = y'^2 : -x'y' : x'^2 \ ; \tag{11}$$

there exists therefore a function F_1 of x, y, x', y' such that

$$F_{x'x'} = y'^2 F_1, \ F_{x'y'} = -x'y'F_1, \ F_{y'y'} = x'^2 F_1 \ . \tag{11a}$$

The function F_1 thus defined is of class C' in the domain $\mathfrak{T}$, even when one of the two variables x', y' is zero; but F_1 becomes in general infinite when x' and y' vanish simultaneously, even if F itself should remain finite and continuous for $x'=0$, $y'=0$.

For instance:

$$F = y\sqrt{x'^2 + y'^2} \ , \qquad F_1 = \frac{y}{(\sqrt{x'^2+y'^2})^3} \ .$$

c) Definition of a Minimum:[1] Two points $A(x_0, y_0)$ and $B(x_1, y_1)$ being given in the region $\mathfrak{R}$, we consider the totality $\mathfrak{M}$ of all ordinary[2] curves which can be drawn in $\mathfrak{R}$ from A to B. Then a curve $\mathfrak{C}$ of $\mathfrak{M}$ is said to minimize the integral

$$J = \int_{t_0}^{t_1} F(x, y, x', y')\, dt \ ,$$

if there exists a neighborhood $\mathfrak{U}$ of $\mathfrak{C}$ such that

$$J_{\mathfrak{C}} \leqq J_{\overline{\mathfrak{C}}} \tag{12}$$

for every ordinary curve $\overline{\mathfrak{C}}$ which can be drawn in $\mathfrak{U}$ from A to B.

We may, without loss of generality, choose for $\mathfrak{U}$ the strip[3] of the x, y-plane swept over by a circle of constant radius ρ whose center moves along the curve $\mathfrak{C}$ from A to B. This strip will be called "the neighborhood (ρ) of $\mathfrak{C}$."

[1] Compare §3. The definition is due to WEIERSTRASS, *Lectures*, 1879; compare also ZERMELO, *Dissertation*, pp. 25–29, and KNESER, *Lehrbuch*, §17.

[2] An extension of the problem to a still more general class of curves will be considered in §31.

[3] In case different portions of the strip should overlap, the plane has to be imagined as multiply covered in the manner of a Riemann-surface (WEIERSTRASS).

§25. THE FIRST VARIATION

We suppose that we have found an ordinary curve

$$\mathfrak{C}: \qquad x=\phi(t)\,, \qquad y=\psi(t)\,, \qquad t_0 \leqq t \leqq t_1\,,$$

contained in the interior of $\mathfrak{R}$, which minimizes the integral J. We replace the curve $\mathfrak{C}$ by a neighboring curve

$$\overline{\mathfrak{C}}: \qquad \bar{x}=x+\xi\,, \qquad \bar{y}=y+\eta\,,$$

where ξ and η are arbitrary functions of t of class D', which vanish at t_0 and t_1:

$$\xi(t_0)=0\,, \quad \eta(t_0)=0\,; \qquad \xi(t_1)=0\,, \quad \eta(t_1)=0\,. \tag{13}$$

The consideration of special variations of the form

$$\xi=\epsilon p\,, \qquad \eta=\epsilon q\,, \tag{14}$$

where ϵ is a constant, and p and q are functions of t of class D', which are independent of ϵ and vanish at t_0 and t_1, leads as in §4 to the result[1] that

$$\Delta J=\delta J+\epsilon(\epsilon)\,, \tag{15}$$

where (ϵ) is an infinitesimal and

$$\delta J=\int_{t_0}^{t_1}(F_x\xi+F_y\eta+F_{x'}\xi'+F_{y'}\eta')\,dt\,, \tag{15a}$$

whence we infer again that δJ must vanish for all admis sible functions ξ, η.

Considering first special variations for which $\eta\equiv 0$, and secondly special variations for which $\xi\equiv 0$, we see that we must have separately

$$\int_{t_0}^{t_1}(F_x\xi+F_{x'}\xi')\,dt=0\,, \quad \int_{t_0}^{t_1}(F_y\eta+F_{y'}\eta')\,dt=0\,. \tag{16}$$

[1] The same results hold for variations of the more general type

$$\xi=\xi(t,\epsilon)\,, \qquad \eta=\eta(t,\epsilon)\,,$$

where the functions $\xi(t,\epsilon)$, $\eta(t,\epsilon)$, their first partial derivatives and the cross-derivatives $\xi_{t\epsilon}$, $\eta_{t\epsilon}$ are continuous in the domain $t_0 \leqq t \leqq t_1$, $|\epsilon| \leqq \epsilon_0$, ϵ_0 being a sufficiently small positive quantity. Moreover

$$\xi(t_0,\epsilon)\equiv 0\,, \qquad \eta(t_0,\epsilon)\equiv 0\,,$$
$$\xi(t_1,\epsilon)\equiv 0\,, \qquad \eta(t_1,\epsilon)\equiv 0\,.$$
$$\xi(t,0)\equiv 0\,, \qquad \eta(t,0)\equiv 0\,, \quad \text{in } (t_0\,t_1)\,.$$

Compare §4, *d*).

To these two equations the methods of §§4–9 can be applied with the following results:

a) Weierstrass's form of Euler's equation: The functions x and y must satisfy the two differential equations

$$F_x - \frac{d}{dt} F_{x'} = 0\ , \qquad F_y - \frac{d}{dt} F_{y'} = 0\ ; \tag{17}$$

these two differential equations are however not independent; for, if we carry out the differentiation with respect to t and make use of the relations (10) and (11a) we obtain

$$F_x - \frac{d}{dt} F_{x'} \equiv y' T\ , \qquad F_y - \frac{d}{dt} F_{y'} \equiv -x' T \tag{18}$$

where
$$T \equiv F_{xy'} - F_{yx'} + F_1 (x' y'' - x'' y')\ , \tag{19}$$

x'', y'' denoting the second derivatives of x and y with respect to t. Since x' and y' do not vanish simultaneously (see §24, a)), the two differential equations (17) are equivalent to the *one differential equation*

$$T \equiv F_{xy'} - F_{yx'} + F_1 (x' y'' - x'' y') = 0\ . \tag{I}$$

This is WEIERSTRASS'S *form of* EULER'S *differential equation.*[1] Every curve satisfying (I) will again be called an *extremal.*

The same result can also be derived from a transformation[1] of δJ which will be useful in the sequel.

If we perform in the expression (15a) for δJ the well-known integration by parts, and make use of (18), we obtain

$$\delta J = \Big[\xi F_{x'} + \eta F_{y'} \Big]_{t_0}^{t_1} + \int_{t_0}^{t_1} T w\, dt\ , \tag{15b}$$

where $w = y'\xi - x'\eta$.

[1] WEIERSTRASS, *Lectures;* compare ZERMELO, *Dissertation*, p. 37.
If we introduce the curvature

$$\frac{1}{r} = \frac{x' y'' - x'' y'}{(\sqrt{x'^2 + y'^2})^3}\ ,$$

the differential equation may also be written

$$\frac{1}{r} = \frac{F_{x'y} - F_{xy'}}{F_1 (\sqrt{x'^2 + y'^2})^3}\ . \tag{Ia}$$

The differential equation (I) together with the initial conditions determines the minimizing curve, but not the functions x and y of t. In order to determine the latter, we must add a second equation or differential equation between t, x, y. This additional relation (which is equivalent to some definite choice of the parameter t) must be such that x and y come out as single-valued functions of t of class D' satisfying (3); otherwise it is arbitrary. The best selection depends largely upon the nature of the particular example under consideration (see the examples in §26).

If we add to (I) a finite relation between t, x, y we obtain as the general solution a pair of functions of t containing two constants of integration:

$$x = f(t, \alpha, \beta) , \qquad y = g(t, \alpha, \beta) . \tag{20}$$

The constants α, β together with the unknown values t_0 and t_1 have to be determined from the condition that the curve must pass through the two given points:

$$\begin{aligned} x_0 &= f(t_0, \alpha, \beta) , & y_0 &= g(t_0, \alpha, \beta) , \\ x_1 &= f(t_1, \alpha, \beta) , & y_1 &= g(t_1, \alpha, \beta) . \end{aligned} \tag{21}$$

b) *Extremal through a given point in a given direction:* In order to construct an extremal through a given point $O(a, b)$ of $\mathfrak{R}$ in a given direction of amplitude γ, we select the arc of the curve measured from the given point for the parameter t and have then to solve the simultaneous system

$$T = 0 , \qquad x'^2 + y'^2 = 1 \tag{22}$$

with the initial conditions

$$x = a , \qquad y = b , \qquad x' = \cos\gamma , \qquad y' = \sin\gamma$$

for $t = 0$. Differentiating the second differential equation we obtain the new system

$$\begin{aligned} F_1 (y'x'' - x'y'') &= F_{xy'} - F_{yx'} , \\ x'x'' + y'y'' &= 0 . \end{aligned} \tag{22a}$$

Solving with respect to x'', y'' we obtain x'', y'' expressed as functions of x, y, x', y' which are of class C' in the vicinity of $x=a$, $y=b$, $x'=\cos\gamma$, $y'=\sin\gamma$ provided that

$$F_1(a, b, \cos\gamma, \sin\gamma) \neq 0 \ . \tag{23}$$

Hence[1] there exists a unique solution

$$x = \Phi(t;\, a, b, \gamma)\,, \qquad y = \Psi(t;\, a, b, \gamma)$$

of the system (22a) satisfying the initial conditions and of class C' in the vicinity of $t=0$.

This solution satisfies also the original system (22). For, by integrating the second equation of (22a) we get: $x'^2+y'^2=$ const., and the value of this constant is found to be 1 from the particular value $t=0$. Thus we reach the result:[2]

If $$F_1(a, b, \cos\gamma, \sin\gamma) \neq 0$$

one and but one extremal of class C' can be drawn through the point (a, b) in the direction γ.

Hence, if (23) is satisfied for every value of γ, a unique extremal of class C' can be drawn from O in every direction.

If (23) is satisfied at every point (a, b) of the region $\mathfrak{R}$ for every value of γ, the problem will be called a *regular problem* (compare §7, *c*)).

c) "*Discontinuous solutions:*" As in §9, *a*) we infer by the method of partial variation that every "discontinuous solution"[3] must be made up of a finite number of arcs of extremals of class C'.

Furthermore, the method of §9, *b*) applied to the two equations (16) leads to the result:[4]

1 According to CAUCHY'S existence-theorem; compare p. 28, footnote 4.

2 See KNESER, *Lehrbuch*, §§27, 29.

3 *I. e.*, a solution which has a finite number of corners; compare §24, *a*).

4 WEIERSTRASS, *Lectures;* compare also KNESER, *Lehrbuch*, §43.

At a corner $t=t_2$ of the minimizing curve, the two conditions

$$F_{x'}\Big|^{t_2-0} = F_{x'}\Big|^{t_2+0}, \qquad F_{y'}\Big|^{t_2-0} = F_{y'}\Big|^{t_2+0} \tag{24}$$

must be satisfied, i. e., the two functions $F_{x'}$ and $F_{y'}$ must remain continuous even at the corners.

We add here the following corollary, though its proof can be given only later (§28):

At a corner (x_2, y_2) of the minimizing curve, the function

$$F_1(x_2, y_2, \cos\theta, \sin\theta)$$

must vanish for some value of the angle θ.

Hence it follows: *If at every point (x, y) of the region $\mathfrak{R}$*

$$F_1(x, y, \cos\theta, \sin\theta) \neq 0$$

for every value of θ, no "discontinuous solutions" are possible.

§26. EXAMPLES

In applications it is frequently convenient to use one of the two equations (17) instead of (I), especially when F does not contain x or y, in which case one of the two equations (17) yields at once a first integral. It must, however, be borne in mind that each of these two equations contains a foreign solution[1] ($y=$const. and $x=$const. respectively), and that only their combination is equivalent to (I).

a) Example X: To determine for a heavy particle the curve of quickest descent in a vertical plane between two given points ("Brachistochrone"[2]).

[1] This happens, for instance, in Example I:

$$F=y\sqrt{x'^2+y'^2},$$

where a first integral is obtained from (17):

$$\frac{yx'}{\sqrt{x'^2+y'^2}}=a;$$

when $a=0$, $y=0$ is such a foreign solution.

[2] Compare LINDELÖF-MOIGNO, *loc. cit.*, No. 112; PASCAL, *loc. cit.*, §31; KNESER, *Lehrbuch*, p. 37.

If we take the positive y-axis vertically downward and denote by g the constant of gravity, by v_0 the initial velocity, which we suppose different from zero, we have to minimize the integral

$$J = \int_{t_0}^{t_1} \frac{\sqrt{x'^2 + y'^2}\, dt}{\sqrt{y - y_0 + k}} ,$$

where

$$k = \frac{v_0^2}{2g} .$$

The curves are restricted to the region

$$\mathfrak{R}: \qquad y - y_0 + k > 0 .$$

Since $F_x \equiv 0$, we obtain the first integral

$$F_{x'} \equiv \frac{x'}{\sqrt{x'^2 + y'^2}\,\sqrt{y - y_0 + k}} = \alpha . \tag{25}$$

The theorem on discontinuous solutions shows that the constant α must have the same value all along the curve.

If $\alpha = 0$, we obtain $x = \text{const.}$, which is the solution of the problem when the two given points A and B lie in the same vertical line.

If $\alpha \neq 0$, we choose for the parameter t the amplitude of the positive tangent to the curve; then we have the additional relation

$$\frac{x'}{\sqrt{x'^2 + y'^2}} = \cos t ,$$

which reduces (25) to

$$y - y_0 + k = r(1 + \cos 2t) ,$$

where

$$r = \frac{1}{2\alpha^2} .$$

Hence

$$y' = -2r \sin 2t ,$$

and

$$x' = \pm 4r \cos^2 t .$$

If we finally make the substitution

$$2t = \tau - \pi ,$$

we get the result

$$\begin{aligned} x - x_0 + h &= \pm r(\tau - \sin \tau) , \\ y - y_0 + k &= r(1 - \cos \tau) , \end{aligned} \tag{26}$$

h being the second constant of integration. The extremals are therefore *cycloids*[1] generated by a circle of radius r rolling upon the horizontal line $y-y_0+k=0$.

Among this double infinitude of cycloids there exists[2] one and but one which passes through the two given points A and B and has no cusp between A and B, provided only that the co-ordinates of the two given points satisfy the inequalities

$$x_1 \neq x_0 \ , \qquad y_1 - y_0 + k \geqq 0 \ .$$

b) *Example XI: To determine the curve of shortest length which can be drawn on a given surface between two given points.*

If the rectangular co-ordinates x, y, z of a point of the surface are given as functions of two parameters u, v and the curves on the surface are expressed in parameter-representation

$$u=\phi(t) \ , \qquad v=\psi(t) \ , \tag{27}$$

the problem is to minimize the integral

$$J=\int_{t_0}^{t_1} \sqrt{Eu'^2+2Fu'v'+Gv'^2}\,dt \ ,$$

where

$$E=\Sigma x_u^2 \ , \qquad F=\Sigma x_u x_v \ , \qquad G=\Sigma x_v^2 \ ,$$

the summation sign referring to a cyclic permutation of x, y, z.

The curves must be restricted to such a portion $\mathfrak{S}$ of the surface that the correspondence between $\mathfrak{S}$ and its image $\mathfrak{R}$ in the u, v-plane is a one-to-one correspondence. We further suppose that E, F, G are of class C'' in $\mathfrak{R}$ and that $\mathfrak{S}$ is free from singular points, *i. e.*,

$$EG-F^2>0 \ .$$

α) If we use *Weierstrass's form (I) of Euler's equation*, and denote by $\Phi(F)$ the differential expression

[1] This result is due to JOHANN BERNOULLI (1696); see OSTWALD'S *Klassiker*, etc., No. 46, p. 3.

[2] See HEFFTER, "Zum Problem der Brachistochrone," *Zeitschrift für Mathematik und Physik*, Vol. XXXIV (1889), p. 313; BOLZA, "The Determination of the Constants in the Problem of the Brachistochrone," *Bulletin of the American Mathematical Society* (2), Vol. X (1904), p. 185; and E. H. MOORE, "On Doubly Infinite Systems of Directly Similar Convex Arches with Common Base Line," *Bulletin of the American Mathematical Society* (2), Vol. X (1904), p. 337.

$$\Phi(F) \equiv F_{xy'} - F_{yx'} + F_1(x'y'' - x''y') \ ,$$

we obtain easily

$$\Phi\left(\sqrt{Eu'^2 + 2Fu'v' + Gv'^2}\right) = \frac{\Gamma}{\left(\sqrt{Eu'^2 + 2Fu'v' + Gv'^2}\right)^3}, \tag{28}$$

where

$$\begin{aligned} \Gamma = {} & (EG - F^2)(u'v'' - u''v') \\ & + (Eu' + Fv')\left[(F_u - \tfrac{1}{2}E_v)u'^2 + G_u u'v' + \tfrac{1}{2}G_v v'^2\right] \\ & - (Fu' + Gv')\left[\tfrac{1}{2}E_u u'^2 + E_v u'v' + (F_v - \tfrac{1}{2}G_u)v'^2\right] . \end{aligned} \tag{29}$$

The extremals satisfy, therefore, the differential equation[1]

$$\Gamma = 0 \ . \tag{29a}$$

This differential equation admits of a simple geometrical interpretation:

The geodesic curvature of the curve (27) at the point t is given by the expression[2]

$$\frac{1}{\rho_g} = \frac{\Gamma}{\sqrt{EG - F^2}\left(\sqrt{Eu'^2 + 2Fu'v' + Gv'^2}\right)^3} \ . \tag{30}$$

Hence the curve of shortest length has the characteristic property that *its geodesic curvature is constantly zero, i. e.*, it is a *geodesic.*

In passing we notice the relation

$$\Phi\left(\sqrt{Eu'^2 + 2Fu'v' + Gv'^2}\right) = \frac{\sqrt{EG - F^2}}{\rho_g} \ , \tag{28a}$$

which will be useful in the sequel.

β) If instead of (I) we use *the two differential equations* (17) and, moreover, select the arc s for the parameter t, we obtain for the extremals the two differential equations:[3]

$$\begin{aligned} 2\frac{d}{ds}\left(E\frac{du}{ds} + F\frac{dv}{ds}\right) &= E_u\left(\frac{du}{ds}\right)^2 + 2F_u\frac{du}{ds}\frac{dv}{ds} + G_u\left(\frac{dv}{ds}\right)^2, \\ 2\frac{d}{ds}\left(F\frac{du}{ds} + G\frac{dv}{ds}\right) &= E_v\left(\frac{du}{ds}\right)^2 + 2F_v\frac{du}{ds}\frac{dv}{ds} + G_v\left(\frac{dv}{ds}\right)^2. \end{aligned} \tag{31}$$

[1] That (29a) is the differential equation of the geodesics might be taken directly from the treatises on differential geometry: KNOBLAUCH, *Flächentheorie*, p. 140; BIANCHI-LUKAT, *Differentialgeometrie*, p. 154; DARBOUX, *Théorie des Surfaces*, Vol. II, p. 417.

[2] See LAURENT, *Traité d'Analyse*, Vol. VII, p. 132.

For an elementary proof see BOLZA, "Concerning the Isoperimetric Problem on a Given Surface," *Decennial Publications of the University of Chicago*, Vol. IX, p. 13.

[3] Compare KNOBLAUCH, *loc. cit.*, p. 142; BIANCHI, *loc. cit.*, p. 153; DARBOUX, *loc. cit.*, p. 419.

They have likewise a simple geometrical meaning: From the definition of E, F, G it follows that

$$E\frac{du}{ds}+F\frac{dv}{ds}=\sum x_u\frac{dx}{ds}\,,$$

$$F\frac{du}{ds}+G\frac{dv}{ds}=\sum x_v\frac{dx}{ds}\,.$$

Differentiating with respect to s we obtain

$$\frac{d}{ds}\left(E\frac{du}{ds}+F\frac{dv}{ds}\right)=\sum x_u\frac{d^2x}{ds^2}$$

$$+\tfrac{1}{2}E_u\left(\frac{du}{ds}\right)^2+F_u\frac{du}{ds}\frac{dv}{ds}+\tfrac{1}{2}G_u\left(\frac{dv}{ds}\right)^2;$$

hence on account of (31)

$$\sum x_u\frac{d^2x}{ds^2}=0\,,$$

and similarly

$$\sum x_v\frac{d^2x}{ds^2}=0\,.$$

Therefore

$$\frac{d^2x}{ds^2}:\frac{d^2y}{ds^2}:\frac{d^2z}{ds^2}=(y_uz_v-y_vz_u):(z_ux_v-z_vx_u):(x_uy_v-x_vy_u)\,. \tag{32}$$

The geometrical meaning of this proportion is that at every point of the curve *the principal normal coincides with the normal to the surface*, which is another characteristic property of the geodesic lines.

§27. THE SECOND VARIATION

Let

$$\mathfrak{E}_0:\quad \begin{aligned} x&=f(t,\alpha_0,\beta_0)=f(t)\,,\\ y&=g(t,\alpha_0,\beta_0)=g(t)\,,\end{aligned}\qquad t_0\leqq t\leqq t_1\,, \tag{33}$$

represent an extremal of class C''' passing through the two given points A and B, derived from the general solution (20) by giving the constants the particular values $\alpha=\alpha_0$, $\beta=\beta_0$.

We suppose that the functions $f(t,\alpha,\beta)$ and $g(t,\alpha,\beta)$, their first partial derivatives and the following higher derivatives,

$$f_{tt},\,f_{t\alpha},\,f_{t\beta},\,f_{tt\alpha},\,f_{tt\beta}\,;\quad g_{tt},\,g_{t\alpha},\,g_{t\beta},\,g_{tt\alpha},\,g_{tt\beta}\,,$$

are continuous in a domain

$$T_0 \leqq t \leqq T_1 \ , \qquad |\alpha - \alpha_0| \leqq d \ , \qquad |\beta - \beta_0| \leqq d$$

where $T_0 < t_0$, $T_1 > t_1$, and d is a sufficiently small positive quantity.

Then we infer, as in §11, that in case of a minimum the second variation of J must be positive or zero. The second variation is defined by the integral

$$\delta^2 J = \int_{t_0}^{t_1} \delta^2 F dt \ ,$$

where

$$\delta^2 F = F_{xx}\xi^2 + 2F_{xy}\xi\eta + F_{yy}\eta^2 + 2F_{xx'}\xi\xi' + 2F_{yy'}\eta\eta' \\ + 2F_{xy'}\xi\eta' + 2F_{yx'}\eta\xi' + F_{x'x'}\xi'^2 + 2F_{x'y'}\xi'\eta' + F_{y'y'}\eta'^2 \ , \tag{34}$$

the arguments of the partial derivatives of F being

$$x = f(t) \ , \qquad y = g(t) \ , \qquad x' = f'(t) \ , \qquad y' = g'(t) \ .$$

a) WEIERSTRASS'S *Transformation of the second variation:*[1] This transformation proceeds by the following steps:

1. Express $F_{x'x'}$, $F_{x'y'}$, $F_{y'y'}$ in terms of F_1 by means of (11a) and introduce the abbreviations

$$w = y'\xi - x'\eta \ , \\ L = F_{xx'} - y'y''F_1 \ , \qquad N = F_{yy'} - x'x''F_1 \ , \\ M = F_{xy'} + x'y''F_1 = F_{yx'} + y'x''F_1 \ ; \tag{35}$$

the two expressions for M are equal since x and y satisfy the differential equation (I).

We thus obtain

$$\delta^2 F = F_1\left(\frac{dw}{dt}\right)^2 + 2L\xi\xi' + 2M(\xi\eta' + \eta\xi') + 2N\eta\eta' \\ + (F_{xx} - y''^2F_1)\xi^2 + 2(F_{xy} + x''y''F_1)\xi\eta + (F_{yy} - x''^2F_1)\eta^2 \, .$$

2. Observe that

$$2L\xi\xi' + 2M(\xi\eta' + \eta\xi') + 2N\eta\eta' \\ = \frac{d}{dt}\left[L\xi^2 + 2M\xi\eta + N\eta^2\right] - \left[\xi^2\frac{dL}{dt} + 2\xi\eta\frac{dM}{dt} + \eta^2\frac{dN}{dt}\right],$$

[1] WEIERSTRASS, *Lectures*, at least as early as 1872.

and introduce the abbreviations

$$
\begin{aligned}
L_1 &= F_{xx} - y''^2 F_1 - \frac{dL}{dt} , \\
M_1 &= F_{xy} + x''y''F_1 - \frac{dM}{dt} , \\
N_1 &= F_{yy} - x''^2 F_1 - \frac{dN}{dt} .
\end{aligned}
\tag{36}
$$

Then the above expression for $\delta^2 F$ becomes

$$
\delta^2 F = F_1 \left(\frac{dw}{dt}\right)^2 + L_1 \xi^2 + 2M_1 \xi\eta + N_1 \eta^2 \\
+ \frac{d}{dt}[L\xi^2 + 2M\xi\eta + N\eta^2] .
$$

3. The three functions L_1, M_1, N_1 have the important property of being proportional to y'^2, $-x'y'$, x'^2.

Proof: From the definition of L, M, N and the relations (10) follows

$$
Lx' + My' = F_x , \qquad Mx' + Ny' = F_y .
$$

Differentiating the first of these relations we get

$$
\frac{dL}{dt}x' + \frac{dM}{dt}y' + Lx'' + My'' \\
= F_{xx}x' + F_{xy}y' + F_{xx'}x'' + F_{xy'}y'' .
$$

But

$$
Lx'' + My'' = F_{xx'}x'' + F_{yx'}y'' ,
$$

and from (I) it follows that

$$
F_{yx'} - F_{xy'} = F_1(x'y'' - x''y') .
$$

Substituting these values we obtain

$$
L_1 x' + M_1 y' = 0 ;
$$

similarly

$$
M_1 x' + N_1 y' = 0 ;
$$

whence we infer that indeed

$$
L_1 : M_1 : N_1 = y'^2 : -x'y' : x'^2 .
$$

There exists therefore a function F_2 of t such that

$$
L_1 = y'^2 F_2 , \qquad M_1 = -x'y'F_2 , \qquad N_1 = x'^2 F_2 . \tag{37}
$$

This reduces the expression for $\delta^2 J$ to the final form

$$\delta^2 J = \int_{t_0}^{t_1} \left[F_1 \left(\frac{dw}{dt} \right)^2 + F_2 w^2 \right] dt + \left[L\xi^2 + 2M\xi\eta + N\eta^2 \right]_{t_0}^{t_1} \qquad (38)$$

If, as we suppose for the present, the two end-points are fixed, then ξ and η vanish at t_0 and t_1 and the expression for $\delta^2 J$ reduces to

$$\delta^2 J = \int_{t_0}^{t_1} \left[F_1 \left(\frac{dw}{dt} \right)^2 + F_2 w^2 \right] dt \ . \qquad (39)$$

This definite integral must then be $\geqq 0$, for all functions w of class D' which vanish at both end-points.

From the assumptions made at the beginning of this section with respect to the functions $f(t, \alpha, \beta)$, $g(t, \alpha, \beta)$ together with our assumptions concerning the function F (see §24, *b*)), it follows that F_1 and F_2 are of class C' in the interval $(T_0 T_1)$; we suppose that they are not both identically zero.

b) WEIERSTRASS'S *form of* LEGENDRE'S *and* JACOBI'S *conditions:* The second variation being now exactly of the same form as in the previous problem (§11), we can directly apply the results of Chapter II.

Accordingly we infer in the first place, as in §11:

The second necessary condition for a minimum (maximum) is that

$$F_1 \geqq 0 \qquad (F_1 \leqq 0) \qquad \text{(II)}$$

along the curve $\mathfrak{C}_0$.

We suppose in the sequel that this condition is satisfied in the slightly stronger form

$$F_1 > 0 \ , \qquad \text{along } \mathfrak{C}_0 \ . \qquad \text{(II')}$$

Again, Jacobi's differential equation (equation (9) of §11) becomes

$$\Psi(u) \equiv F_2 u - \frac{d}{dt}\left(F_1 \frac{du}{dt} \right) = 0 \ . \qquad (40)$$

Jacobi's theorem concerning the integration of this differential equation takes now a slightly different form. If we substitute in the differential equation

$$F_x - \frac{d}{dt} F_{x'} = 0$$

for x and y the general solution

$$x = f(t, \alpha, \beta) \ , \qquad y = g(t, \alpha, \beta)$$

and differentiate with respect to α we get

$$F_{xx} f_\alpha + F_{xy} g_\alpha + F_{xx'} f_{t\alpha} + F_{xy'} g_{t\alpha}$$
$$- \frac{d}{dt} (F_{x'x} f_\alpha + F_{x'y} g_\alpha + F_{x'x'} f_{t\alpha} + F_{x'y'} g_{t\alpha}) = 0 \ .$$

In this equation we express the second partial derivatives of F in terms of L, M, N, F_1, F_2 by means of (11a), (35), (36), (37) and obtain, after some simple reductions,

$$g_t \left[F_2 \omega - \frac{d}{dt} \left(F_1 \frac{d\omega}{dt} \right) \right] = 0 \ ,$$

where

$$\omega = g_t f_\alpha - f_t g_\alpha \ .$$

If we operate in the same manner upon the differential equation

$$F_y - \frac{d}{dt} F_{y'} = 0 \ ,$$

we obtain

$$- f_t \left[F_2 \omega - \frac{d}{dt} \left(F_1 \frac{d\omega}{dt} \right) \right] = 0 \ .$$

Therefore, since f_t and g_t are not both zero, we find that

$$F_2 \omega - \frac{d}{dt} \left(F_1 \frac{d\omega}{dt} \right) = 0 \ .$$

An analogous result is reached if we differentiate with respect to β. Finally, giving α, β the particular values α_0, β_0, we obtain Weierstrass's modification of Jacobi's theorem:

The differential equation

$$\Psi(u) \equiv F_2 u - \frac{d}{dt}\left(F_1 \frac{du}{dt}\right) = 0$$

has the two particular integrals

$$\begin{aligned}\theta_1(t) &= g_t(t) f_\alpha(t) - f_t(t) g_\alpha(t) \ , \\ \theta_2(t) &= g_t(t) f_\beta(t) - f_t(t) g_\beta(t) \ , \end{aligned} \qquad (41)$$

which are in general linearly independent.

Reasoning now as in §§14 and 16 we obtain the result: Let

$$\Theta(t, t_0) = \theta_1(t)\theta_2(t_0) - \theta_2(t)\theta_1(t_0) \ ; \qquad (42)$$

then JACOBI'S *condition* takes the following form:[1]

The third necessary condition for an extremum is that

$$\Theta(t, t_0) \neq 0 \qquad \text{for } t_0 < t < t_1 \ . \qquad \text{(III)}$$

If we denote by t_0' the zero next greater than t_0 of the equation

$$\Theta(t, t_0) = 0 \ ,$$

Condition (III) may also be written:

$$t_1 \leqq t_0' \ ;$$

t_0' is the parameter of the "*conjugate point*" to the point A.

EXAMPLE X[2] (see p. 126).

We suppose that the two end-points A and B lie between the two consecutive cusps $\tau = 0$ and $\tau = 2\pi$ of the cycloid (26), so that the values $\tau = \tau_0$ and $\tau = \tau_1$ corresponding to A and B respectively, satisfy the inequality

$$0 < \tau_0 < \tau_1 < 2\pi \ .$$

For the function F_1 we obtain

$$F_1 = \frac{1}{\sqrt{y - y_0 + k}\left(\sqrt{x'^2 + y'^2}\right)^3} = \frac{1}{8\sqrt{2}\, r^3 \sqrt{r} \sin^4 \frac{\tau}{2}} \ .$$

Hence F_1 is indeed positive along the arc AB.

[1] WEIERSTRASS, *Lectures;* compare also KNESER, *Lehrbuch*, §31.

[2] LINDELÖF-MOIGNO, *loc. cit.*, p. 231.

Again, we obtain from (26)

$$\Theta(\tau, \tau_0) = \pm 4r^2 \sin\frac{\tau}{2}\cos\frac{\tau}{2}\sin\frac{\tau_0}{2}\cos\frac{\tau_0}{2}$$
$$\left[\tau - 2\tan\frac{\tau}{2} - \tau_0 + 2\tan\frac{\tau_0}{2}\right].$$

The parameter τ_0' of the conjugate point A' is therefore determined by the transcendental equation

$$\tau - 2\tan\frac{\tau}{2} = \tau_0 - 2\tan\frac{\tau_0}{2},$$

As τ increases from 0 to π and then from π to 2π, the function $\tau - 2\tan\frac{\tau}{2}$ decreases continually from 0 to $-\infty$ and then from $+\infty$ to $+2\pi$. Hence $\tau = \tau_0$ is the only root of the equation between 0 and 2π. There exists, therefore, *no conjugate point* on the arc AB.

c) KNESER's *form of Jacobi's condition*: As in §15 the existence of a set of extremals through the point A can be proved,[1] representable in the form

$$x = \phi(t, a), \qquad y = \psi(t, a), \tag{43}$$

[1] WEIERSTRASS obtains the set of extremals through A as follows (*Lectures*, 1882): Let

$$\mathfrak{E}: \qquad \bar{x} = f(t, \alpha, \beta), \qquad \bar{y} = g(t, \alpha, \beta)$$

represent the extremal passing through A and making at A a given small angle ω with the extremal

$$\mathfrak{E}_0: \qquad x = f(t, \alpha_0, \beta_0), \qquad y = g(t, \alpha_0, \beta_0).$$

Let further t^0 denote that value of t which corresponds on $\mathfrak{E}$ to the point A. Then we have for the determination of t^0, α, β the three equations:

$$f(t^0, \alpha, \beta) - x_0 = 0, \quad g(t^0, \alpha, \beta) - y_0 = 0, \quad \frac{(x'\bar{y}' - y'\bar{x}')\sqrt{x'^2 + y'^2}}{\sqrt{\bar{x}'^2 + \bar{y}'^2}} - a = 0,$$

where the argument of x', y' is t_0, that of $\bar{x}', \bar{y}'$: t^0, and where

$$a = (x'^2 + y'^2)\sin\omega.$$

The three equations are satisfied for $t^0 = t_0, \alpha = \alpha_0, \beta = \beta_0$; the functions on the left-hand side are continuous and have continuous partial derivatives in the vicinity of $t^0 = t_0, \alpha = \alpha_0, \beta = \beta_0$, and their Jacobian with respect to t^0, α, β is different from zero at this point, since it is equal to

$$\theta_1(t_0)\theta_2'(t_0) - \theta_2(t_0)\theta_1'(t_0),$$

which is different from zero if, as we suppose, $\theta_1(t)$ and $\theta_2(t)$ are linearly independent.

There exists, therefore, according to the theorem on implicit functions, a unique solution t^0, α, β of the above equations, which leads to two functions $\phi(t, a), \psi(t, a)$ having the properties stated in the text.

where $\phi(t, a)$ and $\psi(t, a)$ are continuous with continuous partial derivatives of the first and second orders—with the possible exception of ϕ_{aa}, ψ_{aa}—in the domain

$$T_0 \leqq t \leqq T_1 \ , \qquad |a - a_0| \leqq d_0 \ ,$$

a_0 being the value of a which corresponds to the extremal $\mathfrak{E}_0$ through A and B, and d_0 being a sufficiently small positive quantity.

Again, the Jacobian

$$\frac{\partial(\phi, \psi)}{\partial(t, a)} = \Delta(t, a)$$

differs[1] for $a = a_0$ from the function $\Theta(t, t_0)$ only by a constant factor:

$$\Delta(t, a_0) = C \cdot \Theta(t, t_0) \ , \tag{44}$$

where $C \neq 0$.

Furthermore the value $t = t^0$ which corresponds on the extremal (43) to the point A, and which satisfies therefore the equations

$$x_0 = \phi(t^0, a) \ , \qquad y_0 = \psi(t^0, a) \ , \tag{45}$$

is a function of a, which is, in the vicinity of a_0, of class C'.

From (44) follows KNESER's[2] *form of Jacobi's condition:*

$$\Delta(t, a_0) \neq 0 \qquad \text{for } t_0 < t < t_1 \tag{III}$$

Further, if t_0' denotes the value of t corresponding to the conjugate point A', we have

$$\Delta(t_0', a_0) = 0 \ , \tag{46}$$

and at the same time

$$\Delta_t(t_0', a_0) \neq 0 \ , \tag{47}$$

provided that F_1, F_2 are of class C' in the vicinity of t_0' and $F_1 \neq 0$ at t_0'. The inequality (47) follows[3] from the fact that $\Delta(t, a_0)$ is an integral of Jacobi's differential equation (40).

From this second form of Jacobi's condition it follows[4] easily that the *conjugate point* A' has the *same geometrical meaning* as in the simpler case of §15.

[1] This follows either by direct computation from the equations which define t^0, α, β as functions of a, or else from the fact that $\Delta(t, a_0)$ and $\Theta(t, t_0)$ are integrals of Jacobi's differential equation and vanish for $t = t_0$.

[2] See KNESER, *Lehrbuch*, §31.

[3] Compare p. 58, footnote 2.

[4] See KNESER, *Lehrbuch*, §24, and the references given in E. III D, p. 48, footnote 117.

§28. THE FOURTH NECESSARY CONDITION AND SUFFICIENT CONDITIONS

We suppose in the sequel that for our extremal $\mathfrak{E}_0$ the conditions

$$F_1 > 0 \tag{II'}$$

and

$$\Theta(t, t_0) \neq 0 \qquad \text{for } t_0 < t \leqq t_1 , \tag{III'}$$

are fulfilled.

a) These conditions are not yet sufficient for a (strong) minimum; a *fourth condition* must be added.

Let $\mathbf{E}(x, y;\, x', y';\, \tilde{x}', \tilde{y}')$ be defined[1] as the following function of six independent variables:

$$\begin{aligned}\mathbf{E}(x, y;\, x', y';\, \tilde{x}', \tilde{y}') &= F(x, y, \tilde{x}', \tilde{y}') \\ &\quad - \left[\tilde{x}' F_{x'}(x, y, x', y') + \tilde{y}' F_{y'}(x, y, x', y')\right] ,\end{aligned} \tag{48}$$

or, as we may write on account of (9),

$$\begin{aligned}&\mathbf{E}(x, y;\, x', y';\, \tilde{x}', \tilde{y}') = \\ &\quad \tilde{x}' \left[F_{x'}(x, y, \tilde{x}', \tilde{y}') - F_{x'}(x, y, x', y')\right] \\ &\qquad + \tilde{y}' \left[F_{y'}(x, y, \tilde{x}', \tilde{y}') - F_{y'}(x, y, x', y')\right] .\end{aligned} \tag{48a}$$

Let further (x, y) be any point of the extremal $\mathfrak{E}_0$, p, q the direction-cosines of the positive tangent to $\mathfrak{E}_0$ at (x, y), and $\tilde{p}$, $\tilde{q}$ the direction-cosines of a n y direction.

Then the *fourth necessary condition for a minimum (maximum) is that*

$$\mathbf{E}(x, y;\, p, q;\, \tilde{p}, \tilde{q}) \geqq 0 \qquad (\leqq 0) \tag{IV}$$

for every point (x, y) *of* $\mathfrak{E}_0$ *and for every direction* $\tilde{p}$, $\tilde{q}$.

The proof follows[2] immediately from *Weierstrass's lemma*[3] *on a special class of variations:*

Let

[1] This is WEIERSTRASS's original definition; KNESER writes $-\mathbf{E}$ instead of Weierstrass's $+\mathbf{E}$, *Lehrbuch*, p. 75.

[2] Compare §18, *b*).

[3] The reasoning is the same as in §8; compare also §4, *d*).

$$\mathfrak{C}: \qquad x=\phi(t)\ , \qquad y=\psi(t)\ , \qquad t'\leqq t\leqq t''\ ,$$

be any extremal of class C'' lying in the interior of the region $\mathfrak{R}$, and let $2:(t=t_2)$ be an arbitrary point of $\mathfrak{C}$. Through the point 2 draw an arbitrary curve of class C':

$$\tilde{\mathfrak{C}}: \qquad \tilde{x}=\tilde{\phi}(\tau)\ , \qquad \tilde{y}=\tilde{\psi}(\tau)\ ,$$

the value of $\tau=\tau_2$ corresponding to the point 2.

Let $3:(x_2+\xi_2,\ y_2+\eta_2)$ be the point of $\tilde{\mathfrak{C}}$ corresponding to $\tau=\tau_2+h$, where h is a sufficiently small positive quantity. Finally, from a point $0:(t=t_0<t_2)$ of $\mathfrak{C}$ to the point 3 draw a curve $\bar{\mathfrak{C}}$ representable in the form

$$\bar{\mathfrak{C}}: \qquad \bar{x}=x+\xi\ , \qquad \bar{y}=y+\eta\ ,$$

where ξ and η are functions of t and h which vanish identically for $h=0$, and which satisfy the following conditions[1]:

1. ξ, η themselves, their first partial derivatives and the cross derivatives ξ_{th}, η_{th}, are continuous in the domain

$$t_0\leqq t\leqq t_2\ , \qquad |h|\leqq h_0\ ,$$

h_0 being a sufficiently small positive quantity.

2. $$\xi(t_0,h)=0\ , \quad \eta(t_0,h)=0\ ,$$
$$\xi(t_2,h)=\xi_2\ , \quad \eta(t_2,h)=\eta_2\ ,$$

for every $0\leqq h\leqq h_0$. Then the difference[2]

$$\bar{J}_{03}-(J_{02}+\tilde{J}_{23})$$

has the following value:

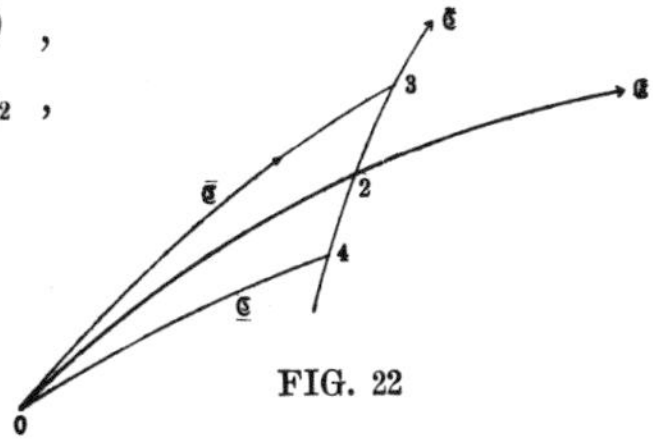

FIG. 22

$$\bar{J}_{03}-(J_{02}+\tilde{J}_{23})=-h\Big[\mathbf{E}(x_2,y_2;\,x_2',y_2';\,\tilde{x}_2',\tilde{y}_2')+(h)\Big]\ . \qquad (49)$$

Similarly, if we denote by 4 the point of $\tilde{\mathfrak{C}}$ corresponding to

[1] Functions ξ, η satisfying these conditions are, for instance, the following:

$$\xi=\xi_2 u\ , \qquad \eta=\eta_2 v\ ,$$

if u, v are two functions of t of class C' which vanish for $t=t_0$ and are equal to 1 for $t=t_2$.

[2] For the notation compare §§2, f), 24 a), and 8.

$\tau = \tau_2 - h$ and draw a curve $\underline{\mathfrak{C}}$ from 0 to 4 of the same character as $\bar{\mathfrak{C}}$, we obtain:

$$\underline{J}_{04} + \tilde{J}_{42} - J_{02} = + h\Big[\mathbf{E}(x_2, y_2;\ x_2', y_2';\ \tilde{x}_2', \tilde{y}_2') + (h)\Big]. \quad (49a)$$

By the same method and under analogous assumptions we further obtain the following results, which are sufficiently explained by the adjoining diagram:

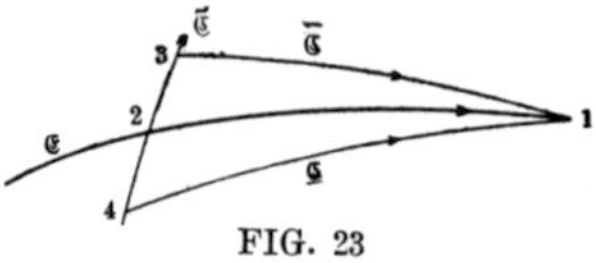

FIG. 23

$$\tilde{J}_{23} + \bar{J}_{31} - J_{21} = h\Big[\mathbf{E}(x_2, y_2;\ x_2', y_2';\ \tilde{x}_2', \tilde{y}_2') + (h)\Big], \quad (50)$$

$$\underline{J}_{41} - (\tilde{J}_{42} + J_{21}) = - h\Big[\mathbf{E}(x_2, y_2;\ x_2', y_2';\ \tilde{x}_2', \tilde{y}_2') + (h)\Big]. \quad (50a)$$

From the relation (8) it follows that

$$\mathbf{E}(x, y;\ kx', ky';\ \tilde{k}\tilde{x}', \tilde{k}\tilde{y}') = \tilde{k}\,\mathbf{E}(x, y;\ x', y';\ \tilde{x}', \tilde{y}'), \quad (51)$$

if $k > 0$ and $\tilde{k} > 0$.

Hence if we set

$$\begin{aligned} p &= \frac{x'}{\sqrt{x'^2 + y'^2}} = \cos\theta, & q &= \frac{y'}{\sqrt{x'^2 + y'^2}} = \sin\theta, \\ \tilde{p} &= \frac{\tilde{x}'}{\sqrt{\tilde{x}'^2 + \tilde{y}'^2}} = \cos\tilde{\theta}, & \tilde{q} &= \frac{\tilde{y}'}{\sqrt{\tilde{x}'^2 + \tilde{y}'^2}} = \sin\tilde{\theta}, \end{aligned} \quad (52)$$

we get

$$\mathbf{E}(x, y;\ x', y';\ \tilde{x}', \tilde{y}') = \sqrt{\tilde{x}'^2 + \tilde{y}'^2}\,\mathbf{E}(x, y;\ p, q;\ \tilde{p}, \tilde{q}), \quad (53)$$

which reduces the second and the third pair of arguments of the **E**-function to direction-cosines.

If we choose for the parameter τ on the curve $\tilde{\mathfrak{C}}$ the arc, we may replace in the above formulae x_2', y_2' and $\tilde{x}_2'$, $\tilde{y}_2'$ by the direction-cosines p_2, q_2 and $\tilde{p}_2$, $\tilde{q}_2$ of the positive tangents at 2 to $\mathfrak{C}$ and to $\tilde{\mathfrak{C}}$ respectively.

b) Relation between the **E**-*function and the function* F_1*:* If the angles θ and $\tilde{\theta}$ are defined by (52), we have, according to (48),

$$\begin{aligned}\mathbf{E}(x, y; p, q; \tilde{p}, \tilde{q}) \\ &= \cos\tilde{\theta}\left[F_{x'}(x, y, \cos\tilde{\theta}, \sin\tilde{\theta}) - F_{x'}(x, y, \cos\theta, \sin\theta)\right] \\ &+ \sin\tilde{\theta}\left[F_{y'}(x, y, \cos\tilde{\theta}, \sin\tilde{\theta}) - F_{y'}(x, y, \cos\theta, \sin\theta)\right] .\end{aligned}$$

But

$$\begin{aligned}F_{x'}(x, y, \cos\tilde{\theta}, \sin\tilde{\theta}) - F_{x'}(x, y, \cos\theta, \sin\theta) \\ &= \int_0^{\omega} \frac{d}{d\tau} F_{x'}\big(x, y, \cos(\theta+\tau), \sin(\theta+\tau)\big)\, d\tau ,\end{aligned}$$

where $\omega = \tilde{\theta} - \theta$; and an analogous formula holds for $F_{y'}$.

If we perform the differentiation with respect to τ and then make use of the relations (11a), we get

$$\begin{aligned}\mathbf{E}(x, y; p, q; \tilde{p}, \tilde{q}) \\ &= \int_0^{\omega} F_1\big(x, y, \cos(\theta+\tau), \sin(\theta+\tau)\big)\sin(\omega-\tau)\, d\tau .\end{aligned}$$

By adding to $\tilde{\theta}$ a proper multiple of 2π, we can always cause ω to lie in the interval

$$-\pi < \omega \leqq \pi ,$$

so that $\sin(\omega - \tau)$ does not change sign between the limits of integration. We may then apply the first mean-value theorem and obtain the following *relation*[1] *between the* **E**-*function and the function* F_1:

$$\begin{aligned}\mathbf{E}(x, y; \cos\theta, \sin\theta; \cos\tilde{\theta}, \sin\tilde{\theta}) \\ &= \big(1 - \cos(\tilde{\theta} - \theta)\big) F_1(x, y, \cos\theta^*, \sin\theta^*) , \qquad (54)\end{aligned}$$

where θ^* is a mean value between θ and $\tilde{\theta}$.

From this theorem follow a number of important consequences:

1. If we let $\tilde{\theta}$ approach θ, we obtain

$$\underset{\tilde{\theta}=\theta}{L} \frac{\mathbf{E}(x, y; p, q; \tilde{p}, \tilde{q})}{1 - \cos(\tilde{\theta} - \theta)} = F_1(x, y, p, q) . \qquad (55)$$

Hence it follows that *Condition* (*II*) *is contained in Condition* (*IV*).

[1] WEIERSTRASS, *Lectures*, 1882.

2. Condition (IV) is always satisfied when

$$F_1(x, y, \cos\gamma, \sin\gamma) \geqq 0 \tag{IIa}$$

for every point (x, y) on $\mathfrak{C}_0$ and for every value of γ.

3. The E-function vanishes whenever $\tilde{\theta} = \theta + 2m\pi$ where m is an integer ("ordinary vanishing")[1]; for a value $\tilde{\theta} \neq \theta + 2m\pi$ where m is an integer it can only vanish[2] ("extra ordinary vanishing") if $F_1(x, y, \cos\gamma, \sin\gamma)$ vanishes for some value $\gamma = \theta^*$ between θ and $\tilde{\theta}$.

c) EXAMPLE XII:[3] To minimize the integral

$$J = \int_{t_0}^{t_1} \frac{yy'^3\,dt}{x'^2 + y'^2} .$$

The value of the E-function is easily found to be

$$\begin{aligned} \mathbf{E}(x, y; p, q; \tilde{p}, \tilde{q}) &= \frac{y(p\tilde{q} - \tilde{p}q)^2\left[(p^2 - q^2)\tilde{q} + 2pq\tilde{p}\right]}{(p^2 + q^2)^2(\tilde{p}^2 + \tilde{q}^2)} \\ &= y \cdot \sin^2(\tilde{\theta} - \theta)\sin(2\theta + \tilde{\theta}) . \end{aligned}$$

Apart from the exceptional case when both end-points lie on the x-axis, $\mathbf{E}$ can be made negative as well as positive by choosing $\tilde{\theta}$ suitably; and therefore no minimum can take place.

More generally, whenever the homogeneity condition (8) holds not only for positive but also for negative values of k, as happens, for instance, when F is a rational function of x', y', no extremum can—in general—take place.

For in this case (51) holds also for negative values of $\tilde{k}$, so that $\mathbf{E}(x, y; p, q; -\tilde{p}, -\tilde{q}) = -\mathbf{E}(x, y; p, q; +\tilde{p}, +\tilde{q})$. Condition (IV) can therefore be fulfilled only if

$$\mathbf{E}(x, y; p, q; \tilde{p}, \tilde{q}) = 0$$

[1] KNESER's terminology, *Lehrbuch*, p. 78.

[2] Hence follows the corollary on discontinuous solutions stated on p. 126. For from (24) follows

$$\mathbf{E}(x, y; \overset{+}{p}, \overset{+}{q}; \overline{p}, \overline{q}) = 0 .$$

[3] To this definite integral leads NEWTON'S celebrated problem: *To determine the solid of revolution of minimum resistance.* Compare PASCAL, *loc. cit.*, p. 111; KNESER, *Lehrbuch*, §§ 11, 18, 26; the above expression for E was given by WEIERSTRASS (1882).

along $\mathfrak{E}_0$ for every direction $\tilde{p}$, $\tilde{q}$, which, on account of (54), is possible only in the exceptional case when $F_1=0$ along $\mathfrak{E}_0$.

d) Sufficiency of the four preceding conditions:[1] The four conditions which so far have been shown to be *necessary* for a minimum of the integral J, are—apart from certain exceptional cases[2]—also *sufficient.*

Let us suppose

1. That $\mathfrak{E}_0$ (or AB) is an arc of an extremal of class C'' without multiple points, lying wholly in the interior of the region[3] $\mathfrak{R}$; (I′)

2. $F_1(x, y, p, q) > 0$ along[4] $\mathfrak{E}_0$; (II′)

3. The arc $\mathfrak{E}_0$ does not contain the conjugate point A' of the point A. (III′)

4. $\mathbf{E}(x, y;\ p, q;\ \tilde{p}, \tilde{q}) > 0$ along[4] $\mathfrak{E}_0$ (IV′)

for every direction $\tilde{p}$, $\tilde{q}$ different from the direction p, q of the positive tangent to $\mathfrak{E}_0$ at (x, y).

Moreover we retain the assumptions made in §27 concerning the general integral of Euler's differential equation.

We propose to prove that under these circumstances the extremal $\mathfrak{E}_0$ actually minimizes the integral

$$J=\int_{t_0}^{t_1} F(x, y, x', y')\,dt\ .$$

From the assumptions (III′) follows the existence of a field of extremals about the arc $\mathfrak{E}_0$, *i. e.*, there exists[5] a neighbor-

[1] WEIERSTRASS, *Lectures*, 1879 and 1882; ZERMELO, *Dissertation*, pp. 77–94; and KNESER, *Lehrbuch*, §20.

[2] The exceptional cases are

1. $\mathfrak{E}_0$ has multiple points or corners, or meets the boundary of $\mathfrak{R}$;
2. $F_1=0$ at certain points of $\mathfrak{E}_0$;
3. A' coincides with B; this case will be considered in §38.
4. $\mathbf{E}=0$ at points of $\mathfrak{E}_0$ for certain directions $\tilde{p}$, $\tilde{q}$ not coinciding with p, q.

[3] Compare §24, *b*).

[4] That is, for every point (x, y) of $\mathfrak{E}_0$, p, q denoting the direction-cosines of the positive tangent to $\mathfrak{E}_0$ at (x, y).

[5] Compare §19. A sharper formulation and a detailed proof of these statements will be given in §34 in connection with Kneser's theory.

hood (ρ) of $\mathfrak{E}_0$ such that to every point P of (ρ) there can be drawn from the point[1] A a uniquely defined extremal which varies continuously with the position of the point P and coincides with $\mathfrak{E}_0$ when P coincides with B.

Let now

$$\overline{\mathfrak{C}}: \qquad \bar{x} = \bar{\phi}(s) \,, \qquad \bar{y} = \bar{\psi}(s) \,, \qquad s_0 \leqq s \leqq s_1 \,,$$

be any ordinary curve drawn from A to B and lying wholly in the neighborhood (ρ) of $\mathfrak{E}_0$, s denoting the arc of the curve $\overline{\mathfrak{C}}$ measured from some fixed point of $\overline{\mathfrak{C}}$, and let ΔJ denote the total variation

$$\Delta J = J_{\overline{\mathfrak{C}}} - J_{\mathfrak{E}_0} \,.$$

Then a reasoning[2] analogous to that employed in §20 leads to the following expression for ΔJ (Weierstrass's Theorem):

$$\Delta J = \int_{s_0}^{s_1} \mathbf{E}\,(\bar{x}, \bar{y}\,; p, q\,; \bar{p}, \bar{q})\, ds \,, \tag{56}$$

where $(\bar{x}, \bar{y})$ denotes a point of $\overline{\mathfrak{C}}$, $\bar{p}$, $\bar{q}$ the direction-cosines of the positive tangent to $\overline{\mathfrak{C}}$ at $(\bar{x}, \bar{y})$, and p, q the direction-cosines of the positive tangent to the unique extremal of the field passing through $(\bar{x}, \bar{y})$.

It now only remains to show that, as a consequence of our assumptions (II′) and (IV′), the integrand in (56) is never negative[3] along the curve $\overline{\mathfrak{C}}$.

Let (x, y) be any point of the above defined neighborhood (ρ) of $\mathfrak{E}_0$ and let, as before, p, q denote the direction-cosines of the positive tangent at (x, y) to the unique extremal of the field passing through (x, y), and $\tilde{p}$, $\tilde{q}$ the direction-cosines of any direction $\tilde{\theta}$, and define

[1] Or better from a point $\bar{A}$ in the vicinity of A on the continuation of $\mathfrak{E}_0$ beyond A, as in §19, c).

[2] The lemma of §8 must be replaced by the lemma of §28, a). Other proofs of Weierstrass's theorem will be given in §37 in connection with Kneser's theory.

[3] It is in this last conclusion that the problem in parameter-representation differs essentially from the problem with x as independent variable; compare §22, c).

$$\mathbf{E}_1(x, y; p, q; \tilde{p}, \tilde{q})$$
$$= \begin{cases} \dfrac{\mathbf{E}(x, y; p, q; \tilde{p}, \tilde{q})}{1-(p\tilde{p}+q\tilde{q})}, & \text{when } 1-(p\tilde{p}+q\tilde{q}) \neq 0 , \\ F_1(x, y, p, q) , & \text{when } 1-(p\tilde{p}+q\tilde{q}) = 0 , \\ & \textit{i. e.}, \ \tilde{p}=p , \quad \tilde{q}=q . \end{cases} \tag{57}$$

The direction-cosines p, q are single-valued and continuous[1] functions of x, y in the neighborhood (ρ) of $\mathfrak{E}_0$. Hence it follows, on account of (54), that $\mathbf{E}_1$ is a continuous function of x, y, $\tilde{\theta}$ in the domain

$$(x, y) \quad \text{in } (\rho) , \qquad 0 \leqq \tilde{\theta} \leqq 2\pi ,$$

and since, according to our assumptions (II′) and (IV′), $\mathbf{E}_1$ is positive along $\mathfrak{E}_0$ for every value of $\tilde{\theta}$, it follows from general theorems on continuous functions that $\mathbf{E}_1$ is positive throughout the domain

$$(x, y) \quad \text{in } (\rho) , \qquad 0 \leqq \tilde{\theta} \leqq 2\pi ,$$

provided that ρ has been taken sufficiently small.

The integrand of (56) is therefore positive at all points of $\bar{\mathfrak{E}}$ at which the direction $\bar{p}$, $\bar{q}$ does not coincide with the direction p, q, and zero where these two directions do coincide. Hence $\Delta J > 0$ unless it should happen that $\bar{p}=p$, $\bar{q}=q$ all along $\bar{\mathfrak{E}}$, in which case we should have $\Delta J = 0$.

But the latter alternative is impossible[2] unless $\bar{\mathfrak{E}}$ be identical with $\mathfrak{E}_0$. This proves that *the arc $\mathfrak{E}_0$ actually minimizes the integral J if the four conditions enumerated at the beginning of §28, d) are fulfilled.*

EXAMPLE VII (see p. 97):

$$F = g(x, y)\sqrt{x'^2 + y'^2} .$$

Here

$$\mathbf{E}_1(x, y; p, q; \tilde{p}, \tilde{q}) = g(x, y) ,$$

[1] Compare §34, Corollary 4.

[2] The proof is similar to that given in §22, *a*); for the details compare KNESER, *Lehrbuch*, §22.

and therefore Condition (IV′) is satisfied if

$$g(x, y) > 0$$

along $\mathfrak{E}_0$.

This shows that in the problem of the brachistochrone an arc AB of the cycloid (26) actually furnishes a minimum if it contains no cusp (compare p. 136).

Corollary: If the condition

$$F_1(x, y, \cos \gamma, \sin \gamma) > 0 \qquad \text{(IIa′)}$$

is satisfied for every point (x, y) of $\mathfrak{E}_0$ and *for every value of* γ, then (II′) and (IV′) are *a fortiori* satisfied, the latter on account of (54).

EXAMPLE XI (see p. 128): *The Geodesics.*
Here

$$F_1 = \frac{E\,G - F^2}{\left(\sqrt{E u'^2 + 2F u' v' + G v'^2}\right)^3} \cdot$$

Hence under the assumptions made on p. 128 concerning the nature of the portion of the surface to which the geodesics are restricted, Condition (IIa′) is always satisfied.

e) Existence of a minimum "im Kleinen": We add here an important theorem which has been used, without proof, by several authors[1] in various investigations of the Calculus of Variations, viz., the theorem that under certain conditions two points can always be joined by a minimizing extremal, provided only that the two points are sufficiently near to each other. An exact formulation and a proof of this theorem have first been given by BLISS.[2] His results are as follows:

We suppose that in addition to our assumptions concerning the function F (see §24, b)) the condition

$$F_1(x, y, \cos \gamma, \sin \gamma) > 0 \qquad (58)$$

---

[1] WEIERSTRASS (*Lectures,* 1879) in his extension of the sufficiency proof to curves without a tangent, see §31; HILBERT in his existence proof (see the references given in chap. vii); OSGOOD in his proof of the identity of Weierstrass's and Hilbert's extension of the meaning of the definite integral J to curves without a tangent (*Transactions of the American Mathematical Society,* Vol. II (1901), p. 295).

[2] *Transactions of the American Mathematical Society,* Vol. V (1904), p. 113. His proof is based upon an extension of a theorem of PICARD's concerning the existence of an integral of a differential equation of the second order, taking for two given values of the independent variable two arbitrarily prescribed values (*Traité d'Analyse,* Vol. III, p. 94).

is fulfilled for every point (x, y) in a finite closed region $\mathfrak{R}_0$ contained in the interior of $\mathfrak{R}$, and for every value of γ.

Since $F_1(x, y, \cos\gamma, \sin\gamma)$ is continuous at every point (x, y) of $\mathfrak{R}$ and for every value of γ, a finite closed region, $\mathfrak{R}_1$, contained in $\mathfrak{R}$ and containing $\mathfrak{R}_0$ in its interior, can be determined such that the inequality (58) still holds for every point (x, y) of $\mathfrak{R}_1$, and for every value of γ.

Under these circumstances, if a positive quantity ϵ be assigned arbitrarily, a second positive quantity ρ_ϵ can be determined such that from every point $P_1(x_1, y_1)$ of $\mathfrak{R}_0$ to every point $P_2(x_2, y_2)$ in the circle (P_1, ρ), where $0<\rho\leqq\rho_\epsilon$, an extremal of class C' can be drawn which lies entirely in the circle (P_1, ρ), and which has the property that at every one of its points the slope with respect to the direction P_1P_2 is numerically less than ϵ. Moreover the circle (P_1, ρ) lies entirely in the region $\mathfrak{R}_1$.

This extremal is at the same time the only extremal of class C' which can be drawn from P_1 to P_2 and which lies entirely in the circle (P_1, ρ).

Let this extremal be represented by

$$\begin{aligned} x &= \Phi(t;\ x_1, y_1;\ x_2, y_2)\ , \\ y &= \Psi(t;\ x_1, y_1;\ x_2, y_2)\ , \end{aligned} \qquad 0\leqq t\leqq t_2\ .$$

Then there exists a positive quantity l, independent of x_1, y_1, x_2, y_2, such that the functions $\Phi, \Psi, \Phi_t, \Psi_t$ are continuous and have continuous first partial derivatives with respect to t, x_1, y_1, x_2, y_2 throughout the domain

$$|t|\leqq l;\ (x_1, y_1)\ \text{ in } \mathfrak{R}_0;\ 0<\sqrt{(x_2-x_1)^2+(y_2-y_1)^2}\leqq\rho\ .$$

Finally also the value $t=t_2$ which corresponds to the point P_2 is a continuous function with continuous first partial derivatives of x_1, y_1, x_2, y_2 for all positions of the two points P_1, P_2 here considered.

For the parameter t of a point P of the extremal we may choose the projection of the vector P_1P upon the vector P_1P_2.

This unique extremal P_1P_2 furnishes for the integral J a smaller value than any other ordinary curve $\overline{\mathfrak{C}}$ which can be drawn from P_1 to P_2 and which lies entirely in the circle (P_1, ρ).

If in addition to the inequality (IIb′) the further condition

$$F(x, y, \cos\gamma, \sin\gamma)>0$$

is fulfilled for every point (x, y) of the region $\mathfrak{R}_0$ and for every value of γ, and if both points P_1 and P_2 lie in $\mathfrak{R}_0$, then the unique extremal P_1P_2 furnishes for the integral J even a smaller value than any ordinary curve, different from the extremal P_1P_2, which can be drawn from P_1 to P_2 and *which lies entirely in* $\mathfrak{R}_0$, provided that $|P_1P_2| \leqq \rho_0$, where ρ_0 is a certain positive quantity less than ρ and independent of the position of P_1 and P_2.

§29. BOUNDARY CONDITIONS[1]

a) Condition along a segment of the boundary: If the minimizing curve 0231 has a segment 23 in common with the boundary of the region $\mathfrak{R}$ to which the admissible curves are confined (see Fig. 7), we obtain the condition which must hold along the boundary as follows:

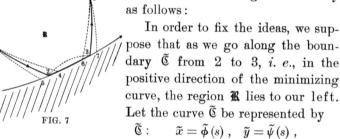

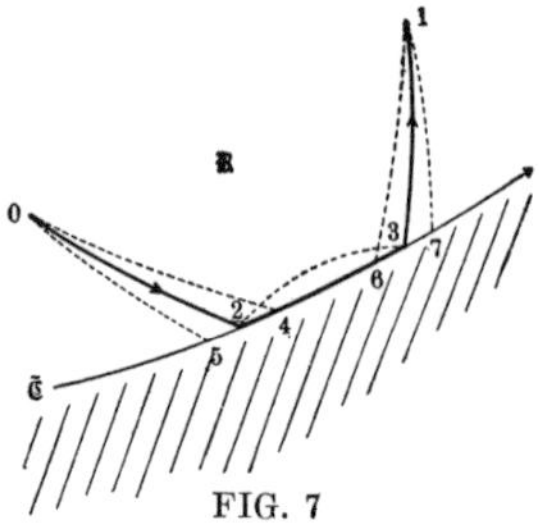

FIG. 7

In order to fix the ideas, we suppose that as we go along the boundary $\tilde{\mathfrak{C}}$ from 2 to 3, *i. e.*, in the positive direction of the minimizing curve, the region $\mathfrak{R}$ lies to our left. Let the curve $\tilde{\mathfrak{C}}$ be represented by

$$\tilde{\mathfrak{C}}: \qquad \tilde{x} = \tilde{\phi}(s), \quad \tilde{y} = \tilde{\psi}(s),$$

s denoting the arc, and suppose that the first and second derivatives of $\tilde{\phi}(s)$ and $\tilde{\psi}(s)$ are continuous along 23.

Then if we construct at a point $(\tilde{x}, \tilde{y})$ of 23 a vector of length u, normal to 23 and directed toward the interior of $\mathfrak{R}$, the co-ordinates of its end-points are

$$\bar{x} = \tilde{x} + \xi, \qquad \bar{y} = \tilde{y} + \eta,$$

where

$$\xi = -\frac{u\tilde{y}'}{\sqrt{\tilde{x}'^2 + \tilde{y}'^2}}, \qquad \eta = \frac{u\tilde{x}'}{\sqrt{\tilde{x}'^2 + \tilde{y}'^2}}.$$

Hence if we substitute for u a function of s of the form

$$u = \epsilon p,$$

[1] Due to WEIERSTRASS, *Lectures*, 1879; compare §10 and KNESER, *Lehrbuch*, §44.

where ϵ is a positive constant and p a function of s of class D' which is $\geqq 0$ in $(s_2 s_3)$ and vanishes at s_2 and s_3, the preceding formulae represent for sufficiently small values of ϵ a curve which remains in the region $\mathfrak{R}$ and which is therefore an admissible variation of the arc 23.

For this variation we obtain, if we apply (15a), for ΔJ the expression

$$\Delta J = \epsilon\left[-\int_{s_2}^{s_3} \tilde{T} p \sqrt{\tilde{x}'^2 + \tilde{y}'^2}\, ds + (\epsilon)\right], \tag{59}$$

from which we infer, by the method of §5, that *in case of a minimum we must have*

$$\tilde{T} \leqq 0 \qquad \text{along } 23\ , \tag{60}$$

where $\tilde{T}$ is the expression (19) in which x, y are replaced by $\tilde{x}$, $\tilde{y}$.

If F_1 is positive not only along the arcs 02 and 31 but also along 23, the preceding condition admits of a simple *geometrical interpretation:*[1] For, if we introduce in the expression for $\tilde{T}$ the curvature $1/\tilde{r}$ of $\tilde{\mathfrak{C}}$ at a point P, and denote by $1/r$ the curvature at the same point P of the extremal which passes through P and is tangent to $\tilde{\mathfrak{C}}$ at P, then (60) may be written, according to equation (Ia) of p. 123, footnote 1,

$$\frac{1}{r} \geqq \frac{1}{\tilde{r}}\ . \tag{61}$$

Hence if $\tilde{r} > 0$, *i. e.*, if the vector from the point P to the center of curvature $\tilde{M}$ of $\tilde{\mathfrak{C}}$ lies to the left of the positive tangent to $\tilde{\mathfrak{C}}$ at P, also r must be positive and the center of curvature M of the extremal must lie between P and $\tilde{M}$ or coincide with $\tilde{M}$.

If, on the contrary, $\tilde{r} < 0$, *i. e.*, if the vector $P\tilde{M}$ lies to the right of the positive tangent, M must lie either on the

[1] This is an extension of the results given for the special case $F = \sqrt{x'^2 + y'^2}$ by KNESER, *Lehrbuch*, p. 178.

opposite side of the tangent to $\tilde{M}$ (when $r > 0$), or else on the same side as, but beyond, $\tilde{M}$ (or coincide with $\tilde{M}$).

If, as we go along the boundary from 2 to 3, the region $\mathfrak{R}$ lies to the right, the condition becomes:

$$\tilde{T} \geqq 0 \qquad \text{along } 23 \tag{60a}$$

or

$$\frac{1}{r} \leqq \frac{1}{\tilde{r}} \ . \tag{61a}$$

b) Conditions at the points of transition: An additional condition must hold at the point 2 where the minimizing curve meets the boundary, and likewise at the point 3 where it leaves the boundary. To obtain the first, let h be a positive infinitesimal and let 4 be the point of $\tilde{\mathfrak{C}}$ whose parameter is $s = s_2 + h$; join the points 0 and 4 by a curve $\bar{\mathfrak{C}}$ of the type defined in §28, a), and consider the variation 0431 of the minimizing curve. For this variation we obtain, according to (49) and (53):

$$\Delta J = \bar{J}_{04} - (J_{02} + \tilde{J}_{24}) = -h\left[\mathbf{E}(x_2, y_2; p_2, q_2; \tilde{p}_2, \tilde{q}_2) + (h)\right] ,$$

where p_2, q_2 and $\tilde{p}_2$, $\tilde{q}_2$ are the direction-cosines of the positive tangents at 2 to the curves 02 and 23 respectively.

Similarly, if we join the point 5 ($s = s_2 - h$) of $\tilde{\mathfrak{C}}$ with the point 0 by a curve $\underline{\mathfrak{C}}$, we get, according to (49a),

$$\Delta J = \underline{J}_{05} + \tilde{J}_{52} - J_{02} = +h\left[\mathbf{E}(x_2, y_2; p_2, q_2; \tilde{p}_2, \tilde{q}_2) + (h)\right] ,$$

whence we infer in the usual manner that *at the point 2 the following condition must be satisfied:*

$$\mathbf{E}(x_2, y_2; p_2, q_2; \tilde{p}_2, \tilde{q}_2) = 0 \ . \tag{62}$$

Applying similar reasoning to the point 3 and making use of (50) and (50a), we reach the result that *at the point 3 the analogous condition*

$$\mathbf{E}(x_3, y_3; p_3, q_3; \tilde{p}_3, \tilde{q}_3) = 0 \tag{63}$$

must be satisfied, where p_3, q_3 and $\tilde{p}_3$, $\tilde{q}_3$ are the direction-cosines of the positive tangents at 3 to 31 and 23 respectively.

The two conditions (62) and (63), together with the condition that the minimizing curve must pass through the given points 0 and 1, determine in general the constants of integration of the two extremals 02 and 31.

If the problem is a "regular" one, *i. e.*, if the condition

$$F_1(x, y, \cos\gamma, \sin\gamma) \neq 0$$

is satisfied at every point (x, y) of the region $\mathfrak{R}$ and for every value of γ, it follows from (54) that (62) and (63) can only be satisfied if

$$\tilde{p}_2 = p_2\,, \quad \tilde{q}_2 = q_2\,; \qquad \tilde{p}_3 = p_3\,, \quad \tilde{q}_3 = q_3\,.$$

This means geometrically that *the arcs 02 and 31 must touch the boundary at the points 2 and 3* in such a manner that their positive tangents coincide with the positive tangents of the boundary.[1]

c) Case where the minimizing curve has only one point in common with the boundary: Suppose that the minimizing curve has only the point 2 in common with the boundary $\tilde{\mathfrak{C}}$. Then the arcs 02 and 21 must be extremals. To find the point 2, let 3 be the point of $\tilde{\mathfrak{C}}$ whose parameter is $s = s_2 + h$, and consider a variation 031 of the curve 021 (see Fig. 24).

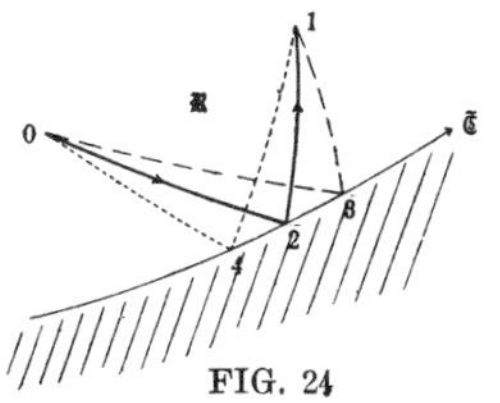

FIG. 24

For this variation we obtain

$$\begin{aligned}\Delta J &= \bar{J}_{03} + \bar{J}_{31} - (J_{02} + J_{21})\\ &= \left[\bar{J}_{03} - (J_{02} + \tilde{J}_{23})\right] + \left[\tilde{J}_{23} + \bar{J}_{31} - J_{21}\right],\end{aligned}$$

which, according to (49) and (49a), is equal to:

$$\begin{aligned}\Delta J = h\Big[&\mathbf{E}(x_2, y_2;\, \overset{+}{p}_2, \overset{+}{q}_2;\, \tilde{p}_2, \tilde{q}_2)\\ &- \mathbf{E}(x_2, y_2;\, \bar{p}_2, \bar{q}_2;\, \tilde{p}_2, \tilde{q}_2) + (h)\Big],\end{aligned}$$

where $\bar{p}_2, \bar{q}_2;\ \overset{+}{p}_2, \overset{+}{q}_2;\ \tilde{p}_2, \tilde{q}_2$ are the direction-cosines of the positive tangents to the arcs 02, 21, 23 respectively at the point 2.

[1] This result is due to ERDMANN, *Journal für Mathematik*, Vol. LXXXII (1877), p. 29.

Similarly, if 4 be the point of $\tilde{\mathfrak{C}}$ whose parameter is $s = s_2 - h$, and we consider a variation 041 of the curve 021, we obtain

$$\begin{aligned}\Delta J &= \left[\bar{J}_{04} - J_{02} + \tilde{J}_{42}\right] + \left[\bar{J}_{41} - (\tilde{J}_{42} + J_{21})\right] \\ &= -h\Big[\mathbf{E}(x_2, y_2; \overset{+}{p}_2, \overset{+}{q}_2; \tilde{p}_2, \tilde{q}_2) \\ &\qquad - \mathbf{E}(x_2, y_2; \bar{p}_2, \bar{q}_2; \tilde{p}_2, \tilde{q}_2) + (h)\Big] .\end{aligned}$$

Hence we infer that *at the point 2 the condition*[2]

$$\mathbf{E}(x_2, y_2; \bar{p}_2, \bar{q}_2; \tilde{p}_2, \tilde{q}_2) = \mathbf{E}(x_2, y_2; \overset{+}{p}_2, \overset{+}{q}_2; \tilde{p}_2, \tilde{q}_2) \tag{64}$$

must be satisfied.

d) EXAMPLE VI[1] (see p. 84):

$$F = \sqrt{x'^2 + y'^2} .$$

Suppose the region $\mathfrak{R}$ to be the whole plane with the exception of the interior of a simply closed curve of class C'', and suppose that the straight line joining 0 and 1 passes through the excluded region.

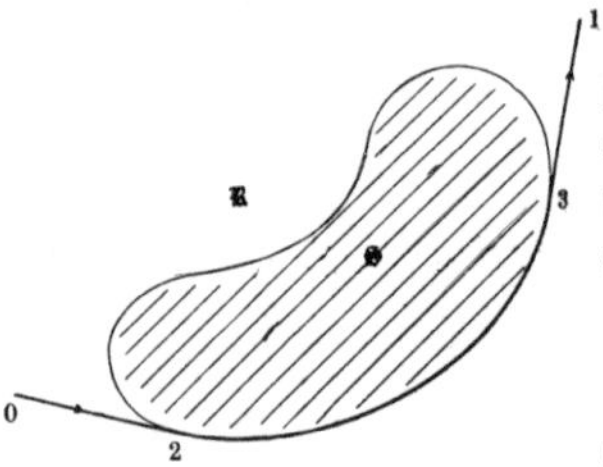

FIG. 25

The minimizing curve must be composed of segments of straight lines and segments of the boundary, the latter turning their *convex side outward* since in this case $1/r = 0$ and therefore

$$\frac{1}{\tilde{r}} \leqq 0 \qquad \text{or} \quad \geqq 0 ,$$

according as 23 is described positively or negatively with respect to $\mathfrak{R}$. The lines 02 and 31 must touch the arc 23 positively at 2 and 3 since $F_1(x, y, \cos\gamma, \sin\gamma) = 1$.

Again,

$$\mathbf{E}(x, y; \cos\theta, \sin\theta; \cos\tilde{\theta}, \sin\tilde{\theta}) = 1 - \cos(\tilde{\theta} - \theta) .$$

Hence if the minimizing curve is to have one point 2 in common with the boundary, the condition

$$\cos(\tilde{\theta}_2 - \bar{\theta}_2) = \cos(\tilde{\theta}_2 - \overset{+}{\theta}_2)$$

must be satisfied at 2. This means that the lines 02 and 21 must make equal angles with the tangent to the boundary at 2.

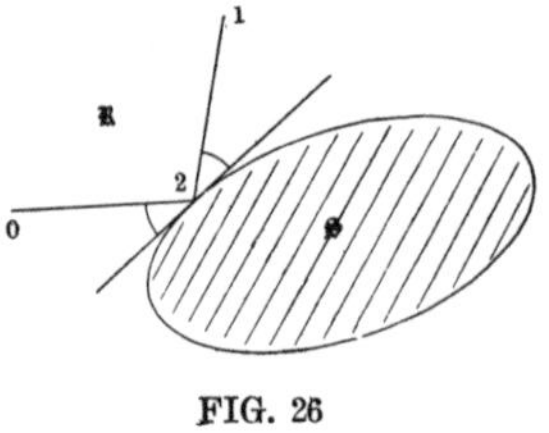

FIG. 26

[1] Compare KNESER, *Lehrbuch*, p. 178. [2] See p. 265.

e) EXAMPLE I (see p. 1):

$$F = y\sqrt{x'^2 + y'^2} \ ;$$

the region $\mathfrak{R}$ is the upper half-plane:

$$y \geqq 0 \ .$$

The extremals are here

α) The catenaries

$$x = t \ , \qquad y = \alpha \cosh \frac{t-\beta}{\alpha} \ ;$$

β) The straight lines

$$x = \alpha, \qquad y = t \ .$$

Since the catenaries never meet the x-axis, the only possible solution containing a segment of the boundary consists of the ordinates of the two given points:

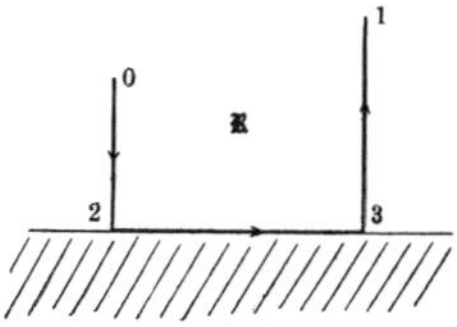

FIG. 27

$$x = x_0 \qquad \text{and} \qquad x = x_1 \ ,$$

together with the segment 23 of the x-axis between them.

Since along the x-axis

$$T = -1 \ ,$$

condition (60) is satisfied along 23; and since

$$\mathbf{E}(x, y;\ \cos\theta, \sin\theta;\ \cos\tilde{\theta}, \sin\tilde{\theta}) = \left(1 - \cos(\tilde{\theta} - \theta)\right) y \ ,$$

conditions (62) and (63) are satisfied at 2 and 3.

§30. THE CASE OF VARIABLE END-POINTS

The methods explained in §23, slightly modified, can be applied to the case when all curves considered are expressed in parameter-representation. In one respect the treatment of the problem in parameter-representation is even considerably simpler, viz.: *the variation of the limits of the integral J can be completely avoided.* For let

$$\mathfrak{C}_0: \qquad x = \phi(t) \ , \qquad y = \psi(t) \ , \qquad t_0 \leqq t \leqq t_1 \ , \qquad (65)$$

be the minimizing curve, and

$$\bar{\mathfrak{C}}: \qquad \bar{x} = \bar{\phi}(\tau) \ , \qquad \bar{y} = \bar{\psi}(\tau) \ , \qquad \tau_0 \leqq \tau \leqq \tau_1 \ , \qquad (66)$$

a neighboring curve. If we then apply to $\bar{\mathfrak{C}}$ the "parameter-transformation" (see §24, a))

$$t = t_0 + \frac{(t_1 - t_0)(\tau - \tau_0)}{\tau_1 - \tau_0}$$

we obtain for $\bar{\mathfrak{C}}$ a representation in terms of the parameter t for which the end-values are t_0 and t_1, the same as for $\mathfrak{E}_0$.

We consider briefly the case where the point 1 is fixed and the point 0 movable on a given curve of class C'':

$$\tilde{\mathfrak{C}}: \qquad \tilde{x} = \tilde{\phi}(a) \ , \qquad \tilde{y} = \tilde{\psi}(a) \ . \tag{67}$$

The minimizing curve (65) must again be an extremal; it begins at a point 0 of the curve $\tilde{\mathfrak{C}}$ whose parameter on $\tilde{\mathfrak{C}}$ we denote by a_0. Let $2: (a = a_0 + \epsilon)$ be a point of $\tilde{\mathfrak{C}}$ in the vicinity of 0, $x_0 + \xi_0$, $y_0 + \eta_0$ its co-ordinates; then

$$\xi_0 = \epsilon\left[\tilde{\phi}'(a_0) + (\epsilon)\right] , \qquad \eta_0 = \epsilon\left[\tilde{\psi}'(a_0) + (\epsilon)\right] .$$

An admissible variation $\bar{\mathfrak{C}}$ of sufficient generality which passes through 2 and 1, can easily be constructed analytically in the form

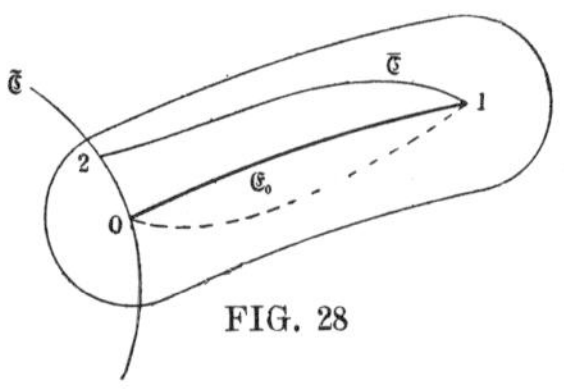

FIG. 28

$$\bar{\mathfrak{C}}: \qquad \bar{x} = x + \xi \ , \qquad \bar{y} = y + \eta \ ,$$

where

$$\xi = \xi_0 u \ , \qquad \eta = \eta_0 v \ ,$$

u, v being two arbitrary functions of t of class C' which vanish for $t = t_1$ and are equal to 1 for $t = t_0$.

For this variation of the curve $\mathfrak{E}_0$ we obtain, according to (15b),

$$\Delta J = \Big[\xi F_{x'} + \eta F_{y'}\Big]_{t_0}^{t_1} + \int_{t_0}^{t_1} (y'\xi - x'\eta)\, T\, dt + \epsilon(\epsilon) \ .$$

Substituting the values of ξ, η at t_0 and t_1 and remembering that $T = 0$ along the extremal $\mathfrak{E}_0$, we get[1]

$$\Delta J = \epsilon\Big[-(\tilde{x}'F_{x'} + \tilde{y}'F_{y'})\Big|^{0} + (\epsilon)\Big] ,$$

where

[1] WEIERSTRASS, *Lectures*, 1882.

$$\tilde{x}'=\frac{d\tilde{x}}{da}\,, \qquad \tilde{y}'=\frac{d\tilde{y}}{da}\,.$$

We obtain, therefore, the *condition of transversality* in the form

$$\tilde{x}'F_{x'}(x,y,x',y')+\tilde{y}'F_{y'}(x,y,x',y')\Big|^0=0 \qquad (68)$$

where x', y' refer to the extremal $\mathfrak{C}_0$, $\tilde{x}'$, $\tilde{y}'$ to the given curve $\tilde{\mathfrak{C}}$.

EXAMPLE XI (see p. 128): *The Geodesics.* The condition of transversality is

$$\tilde{u}'(Eu'+Fv')+\tilde{v}'(Fu'+Gv')=0\;; \qquad (69)$$

its geometrical meaning[1] is that the geodesic must be *orthogonal* to the given curve.

The *focal point* is determined by the following formulae:[2] Let A_0 and B_0 denote the following two constants

$$A_0=\frac{\tilde{x}''F_{x'}+\tilde{y}''F_{y'}+L\tilde{x}'^2+2M\tilde{x}'\tilde{y}'+N\tilde{y}'^2}{\tilde{x}'^2+\tilde{y}'^2}\Bigg|^0,$$
$$B_0=\frac{(x'\tilde{y}'-y'\tilde{x}')^2F_1}{\tilde{x}'^2+\tilde{y}'^2}\Bigg|^0, \qquad (70)$$

where the arguments of $F_{x'}$, $F_{y'}$, F_1 are x_0, y_0, x_0', y_0' and L, M, N are defined by (35). B_0 is different from zero if we suppose, as in §23, that $\mathfrak{C}_0$ and $\tilde{\mathfrak{C}}$ are not tangent to each other at the point 0. Let further

$$\mathrm{H}(t_0,t)=A_0\Theta(t_0,t)+B_0\frac{\partial\Theta(t_0,t)}{\partial t_0}\,, \qquad (71)$$

the function Θ being defined by (42). Then the parameter t_0'' of the focal point is given by the equation

$$\mathrm{H}(t_0,t)=0\,. \qquad (72)$$

If

$$x=\phi(t,a)\,, \qquad y=\psi(t,a)$$

[1] Compare BIANCHI-(LUKAT), *Differentialgeometrie*, p. 65.

[2] See BLISS, *Transactions of the American Mathematical Society*, Vol. III (1902) p. 136.

is the extremal which passes through the point a of the curve $\tilde{\mathfrak{C}}$ and is cut transversely by $\tilde{\mathfrak{C}}$ at that point, and if $\Delta(t, a)$ denotes the Jacobian of the two functions ϕ, ψ with respect to t, a, then[1]

$$\Delta(t, a) = C\mathrm{H}(t_0, t) \tag{73}$$

which proves the geometrical meaning of the focal point.

The question of sufficient conditions will be discussed in detail in connection with Kneser's theory in chap. v.

§31. WEIERSTRASS'S EXTENSION OF THE MEANING OF THE DEFINITE INTEGRAL

$$\int_{t_0}^{t_1} F(x, y, x', y')\, dt\ .$$

We have confined[2] ourselves in all the preceding investigations to "ordinary" curves. This limitation was indeed necessary for most of our proofs, but it is not implied in the nature of the problem.

The most general class of curves for which the problem has a meaning would be the totality of curves for which the integral

$$J = \int_{t_0}^{t_1} F(x, y, x', y')\, dt$$

is finite and determinate.

In many problems of a geometrical origin, however, a still further generalization is desirable.

a) Example of the length of a curve: Thus, for instance, the problem to determine the curve of shortest length between two given points A and B, is not exactly equivalent to the problem to minimize the integral

$$J = \int_{t_0}^{t_1} \sqrt{x'^2 + y'^2}\, dt\ ,$$

because the length of a curve cannot in all cases be expressed by this integral.

The length of a continuous curve

[1] See BLISS, *loc. cit.*, p. 140.

[2] Compare §24, a) and c).

$$\mathfrak{L}: \qquad x = \phi(t) \;, \qquad y = \psi(t) \;, \qquad t_0 \leqq t \leqq t_1 \tag{74}$$

is defined[1] as follows:

Consider any partition Π of the interval $(t_0 t_1)$ into n subintervals by points of division $\tau_1, \tau_2, \cdots, \tau_{n-1}$, where

$$t_0 < \tau_1 < \tau_2 \cdots < \tau_{n-1} < t_1 \;,$$

and denote by $A, P_1, P_2, \cdots, P_{n-1}, B$ the corresponding points of $\mathfrak{L}$, by X_0, Y_0; x_1, y_1; x_2, y_2; $\cdots$; x_{n-1}, y_{n-1}; X_1, Y_1 their co-ordinates. Then the length of the polygon $\mathfrak{P}_\Pi$ inscribed in the curve $\mathfrak{C}$ whose successive vertices are these points, is

$$S_\Pi = \sum_{\nu=0}^{n-1} \sqrt{(\Delta x_\nu)^2 + (\Delta y_\nu)^2} \;,$$

where[2]

$$\Delta x_\nu = x_{\nu+1} - x_\nu \;, \qquad \Delta y_\nu = y_{\nu+1} - y_\nu \;.$$

If S_Π approaches a determinate finite limit[3] J as all the differences $(\tau_{\nu+1} - \tau_\nu)$ approach zero:

$$J = \underset{\Delta\tau=0}{L} S_\Pi \;,$$

the curve $\mathfrak{L}$ is said to have a finite length whose value is J.

If the first derivatives $\phi'(t), \psi'(t)$ exist and are continuous in $(t_0 t_1)$, the above limit always exists and can be expressed by the definite integral[4]

$$\int_{t_0}^{t_1} \sqrt{x'^2 + y'^2}\, dt \;.$$

b) Extension of the meaning of the general integral: In an entirely analogous manner WEIERSTRASS[5] has generalized the meaning of the definite integral

[1] See JORDAN, *Cours d'Analyse*, Vol. I, Nos. 105–111. This is the definition which is most convenient for our present purpose; compare also §44, *a*), end.

[2] With the understanding that $\tau_0 = t_0$, $x_0 = X_0$, $y_0 = Y_0$ and $\tau_n = t_1$, $x_n = X_1$, $y_n = Y_1$.

[3] That is, corresponding to every positive ϵ, another positive quantity δ_ϵ can be assigned such that

$$|J - S_\Pi| < \epsilon$$

for all partitions Π in which all the differences $(\tau_{\nu+1} - \tau_\nu)$ are less than δ_ϵ.

[4] Compare JORDAN, *loc. cit.*, No. 111, and STOLZ, *Transactions of the American Mathematical Society*, Vol. III (1902), pp. 28 and 303.

[5] *Lectures*, 1879; compare also OSGOOD, *Transactions of the American Mathematical Society*, Vol. II (1901), pp. 275 and 293.

$$J = \int_{t_0}^{t_1} F(x, y, x', y')\, dt \ ,$$

taken along a continuous curve $\mathfrak{L}$ (defined by (74)) which lies entirely in the interior of the region $\mathfrak{R}$ of § 24, *b*).

Consider as before a partition Π of the interval $(t_0 t_1)$ and denote by W_Π the sum

$$W_\Pi = \sum_{\nu=0}^{n-1} F(x_\nu, y_\nu, \Delta x_\nu, \Delta y_\nu) \ . \tag{75}$$

Then, if the curve $\mathfrak{L}$ is of class[1] C', this sum W_Π approaches a determinate finite limit as all the differences $(\tau_{\nu+1} - \tau_\nu)$ approach zero, viz., the definite integral[2] $J_{\mathfrak{L}}(AB)$:

$$\underset{\Delta\tau=0}{L} W_\Pi = \int_{t_0}^{t_1} F(x, y, x', y')\, dt \ . \tag{76}$$

This remains true when $\mathfrak{L}$ has a finite number of corners.

We now agree to define the definite integral

$$\int_{t_0}^{t_1} F(x, y, x', y')\, dt \ ,$$

[1] This implies that $\phi'^2(t) + \psi'^2(t) \neq 0$ in $(t_0 t_1)$; compare § 24, *a*).

[2] For the definite integral may be written

$$J = \sum_{\nu=0}^{n-1} \int_{\tau_\nu}^{\tau_{\nu+1}} F(x, y, x', y')\, dt = \sum_{\nu=0}^{n-1} F\big(\phi(\tau'_\nu), \psi(\tau'_\nu), \phi'(\tau'_\nu), \psi'(\tau'_\nu)\big)(\tau_{\nu+1} - \tau_\nu) \ ,$$

where τ'_ν is some intermediate value between τ_ν and $\tau_{\nu+1}$. On the other hand

$$\Delta x_\nu = \phi'(\tau''_\nu)(\tau_{\nu+1} - \tau_\nu) \ , \qquad \Delta y_\nu = \psi'(\tau'''_\nu)(\tau_{\nu+1} - \tau_\nu) \ ,$$

where τ''_ν and τ'''_ν are again intermediate values between τ_ν and $\tau_{\nu+1}$. Hence we have, on account of the homogeneity of F,

$$W_\Pi = \sum_{\nu=0}^{n-1} F\big(\phi(\tau_\nu), \psi(\tau_\nu), \phi'(\tau''_\nu), \psi'(\tau'''_\nu)\big)(\tau_{\nu+1} - \tau_\nu) \ .$$

From the theorem on uniform continuity applied to the function $F(x, y, x', y')$ on the one hand, and to the functions $\phi(t)$, $\psi(t)$ and their derivatives on the other hand, it follows that corresponding to every positive quantity ϵ another positive quantity δ_ϵ can be determined such that

$$\left| F\big(\phi(\tau_\nu), \psi(\tau_\nu), \phi'(\tau''_\nu), \psi'(\tau'''_\nu)\big) - F\big(\phi(\tau'_\nu), \psi(\tau'_\nu), \phi'(\tau'_\nu), \psi'(\tau'_\nu)\big) \right| < \epsilon$$

for $\nu = 0, 1, 2, \cdots, n-1$, provided that all the differences $(\tau_{\nu+1} - \tau_\nu)$ are less than δ_ϵ. Hence

$$| W_\Pi - J | < \epsilon (t_1 - t_0) \ ,$$

which proves our statement.

taken along the curve $\mathfrak{L}$, as the limit of W_{Π} in all cases in which this limit exists and is finite: and we denote its value by $J^*_{\mathfrak{L}}(AB)$:

$$J^*_{\mathfrak{L}}(AB) = \underset{\Delta\tau=0}{L} W_{\Pi} \ . \tag{77}$$

This is a natural extension of the definition of the definite integral since it coincides with the ordinary definition for all "ordinary" curves.

c) First modification of Weierstrass's definition: Various modifications of this definition will be of importance in the sequel:

Since the curve $\mathfrak{L}$ is supposed to lie in the interior of the region $\mathfrak{R}$, the rectilinear polygon whose vertices are the points $A, P_1, P_2, \cdots, P_{n-1}, B$ will likewise lie in the interior of $\mathfrak{R}$, provided that the differences $(\tau_{\nu+1} - \tau_\nu)$ have been taken sufficiently small. Let V_{Π} denote the value of the integral J taken along this polygon from A to B.

If, then, the curve $\mathfrak{L}$ is rectifiable, and if one of the two sums V_{Π} and W_{Π} approaches for $L\Delta\tau = 0$ a determinate finite limit, the other approaches the same limit,[1] so that we may also define

$$J^*_{\mathfrak{L}}(AB) = \underset{\Delta\tau=0}{L} V_{\Pi} \ . \tag{78}$$

d) Second modification of Weierstrass's definition: If the curve $\mathfrak{L}$ is *rectifiable and lies in a finite closed region* $\mathfrak{R}_0$ *(contained in the interior of the region* $\mathfrak{R}$*) in which the condition*

$$F_1(x, y, \cos\gamma, \sin\gamma) > 0 \tag{58}$$

is fulfilled for every value of γ, then the preceding extension of the meaning of the definite integral J may be modified as follows:

Let a positive quantity ϵ be chosen arbitrarily. Then determine for the region $\mathfrak{R}_0$ the quantity ρ_ϵ defined in § 28, *e*) and choose a positive quantity $\rho \leqq \rho_\epsilon$ arbitrarily. Further select, according to

[1] See OSGOOD, *Transactions of the American Mathematical Society*, Vol. II (1901), p. 293. If $l_{\nu+1}$ and $\gamma_{\nu+1}$ denote the length and the amplitude of the vector $P_\nu P_{\nu+1}$, the difference $V_{\Pi} - W_{\Pi}$ may be written in the form

$$V_{\Pi} - W_{\Pi} = \sum_{\nu=0}^{n-1} \int_0^{l_{\nu+1}} \Big[F(\bar{x}_{\nu+1}, \bar{y}_{\nu+1}, \cos\gamma_{\nu+1}, \sin\gamma_{\nu+1}) - F(x_\nu, y_\nu, \cos\gamma_{\nu+1}, \sin\gamma_{\nu+1})\Big] ds\,,$$

where $\bar{x}_{\nu+1} = x_\nu + s\cos\gamma_{\nu+1}$, $\quad \bar{y}_{\nu+1} = y_\nu + s\sin\gamma_{\nu+1}$.

The above statement follows, then, from the theorem on uniform continuity applied to the function $F(x, y, x', y')$.

the theorem on uniform continuity, another positive quantity δ so small that

$$|\phi(t') - \phi(t'')| < \rho/\sqrt{2}\ , \qquad |\psi(t') - \psi(t'')| < \rho/\sqrt{2}$$

for every two values t', t'' of the interval (t_0t_1) for which

$$|t'' - t'| < \delta\ .$$

Finally choose the partition Π so that

$$\tau_{\nu+1} - \tau_\nu < \delta$$

for $\nu = 0, 1, 2, \cdots, n-1$.

Then the distance $|P_\nu P_{\nu+1}|$ is less than ρ, and therefore we can, according to § 28, *e*), *inscribe in the curve* $\mathfrak{L}$ *a unique polygon of minimizing extremals with the points* $A, P_1, P_2, \cdots, P_{n-1}, B$ *for vertices, i. e.*, we can draw from P_ν to $P_{\nu+1}$ a unique extremal $\mathfrak{E}_{\nu+1}$ of class C' which lies entirely in the circle (P_ν, ρ) and which furnishes for the integral J a smaller value than any other ordinary curve which can be drawn from P_ν to $P_{\nu+1}$ and which lies entirely in the circle (P_ν, ρ), Moreover, at every point of $\mathfrak{E}_{\nu+1}$ the slope with respect to the direction $P_\nu P_{\nu+1}$ is less than ϵ.

We denote by U_Π the value of the integral J taken along this polygon of extremals, *i. e.*,

$$U_\Pi = \sum_{\nu=0}^{n-1} J_{\mathfrak{E}_{\nu+1}}(P_\nu P_{\nu+1})\ . \tag{79}$$

Then if we pass, as before, to the limit $L\,\Delta\tau = 0$, and *if one of the two sums* U_Π *and* W_Π *approaches a finite and determinate limit, the other approaches the same limit,*[1] *so that we may also define*

[1] First remarked by OSGOOD, *Transactions of the American Mathematical Society*, Vol. II (1901), p. 295. The statement can be proved as follows:

Let the extremal $\mathfrak{E}_{\nu+1}$ be represented by

$$\mathfrak{E}_{\nu+1}: \qquad x = \phi_{\nu+1}(t)\ , \qquad y = \psi_{\nu+1}(t)\ , \qquad 0 \leqq t \leqq l_{\nu+1}\ ,$$

where, as in § 28, *e*), the parameter t of a point P of $\mathfrak{E}_{\nu+1}$ is the projection $P_\nu Q$ of the vector $P_\nu P$ upon the vector $P_\nu P_{\nu+1}$, and $l_{\nu+1}$ is again the distance $|P_\nu P_{\nu+1}|$. If we denote by $\gamma_{\nu+1}$ the amplitude of the vector $P_\nu P_{\nu+1}$ and by u the perpendicular QP with the sign $+$ or $-$ according as the point P lies to the left or to the right of the vector $P_\nu P_{\nu+1}$, then we have

$$\phi_{\nu+1}(t) = x_\nu + t\cos\gamma_{\nu+1} - u\sin\gamma_{\nu+1}\ , \qquad \psi_{\nu+1}(t) = y_\nu + t\sin\gamma_{\nu+1} + u\cos\gamma_{\nu+1}\ ,$$

$$\phi'_{\nu+1}(t) = \cos\gamma_{\nu+1} - u'\sin\gamma_{\nu+1}\ , \qquad \psi'_{\nu+1}(t) = \sin\gamma_{\nu+1} + u'\cos\gamma_{\nu+1}\ .$$

Hence if we write

$$\phi_{\nu+1}(t) = x_\nu + \xi_\nu\ , \qquad \psi_{\nu+1}(t) = y_\nu + \eta_\nu\ ,$$

$$\phi'_{\nu+1}(t) = \cos\gamma_{\nu+1} + \zeta_\nu\ , \qquad \psi'_{\nu+1}(t) = \sin\gamma_{\nu+1} + \theta_\nu\ ,$$

$$J^*_{\mathfrak{L}}(AB) = \underset{\Delta\tau=0}{L}\, U_{\text{II}} \,. \tag{80}$$

We shall call the totality of rectifiable curves for which the sum W_{II} approaches a determinate finite limit, "*the class* (K)."

e) Extension of the sufficiency proof to curves of class (K)*:* After these preliminaries, let $\mathfrak{E}_0$ denote an *extremal of class* C' *drawn from A to B and lying wholly in the interior of the region* $\mathfrak{R}$. We suppose that $\mathfrak{E}_0$ *does not contain the conjugate A' to the point A*, and that *for every point (x, y) of $\mathfrak{E}_0$ and for every value of γ the condition*

we have for every t in the interval $(0 l_{\nu+1})$

$$|\xi_\nu| \leqq \rho \,, \qquad |\eta_\nu| \leqq \rho \,,$$

since $\mathfrak{E}_{\nu+1}$ lies in the circle (P_ν, ρ); and

$$|\zeta_\nu| < \epsilon \,, \qquad |\theta_\nu| < \epsilon \,,$$

since the slope u' of $\mathfrak{E}_{\nu+1}$ at the point P with respect to the direction $P_\nu P_{\nu+1}$ is numerically less than ϵ.

Applying now to the integral $J_{\mathfrak{E}_{\nu+1}}$ the first mean-value theorem we obtain

$$J_{\mathfrak{E}_{\nu+1}}(P_\nu P_{\nu+1}) = l_{\nu+1} F(x_\nu + \tilde{\xi}_\nu, y_\nu + \tilde{\eta}_\nu, \cos\gamma_{\nu+1} + \tilde{\zeta}_\nu, \sin\gamma_{\nu+1} + \tilde{\theta}_\nu) \,,$$

where the argument of $\tilde{\xi}_\nu, \tilde{\eta}_\nu, \tilde{\zeta}_\nu, \tilde{\theta}_\nu$ is some value of t between 0 and $l_{\nu+1}$.

On the other hand, we have on account of the homogeneity of F,

$$F(x_\nu, y_\nu, \Delta x_\nu, \Delta y_\nu) = l_{\nu+1} F(x_\nu, y_\nu, \cos\gamma_{\nu+1}, \sin\gamma_{\nu+1}) \,.$$

The extremal of $\mathfrak{E}_{\nu+1}$—though it need not lie entirely in the region $\mathfrak{R}_0$—certainly lies in the larger region $\mathfrak{R}_1$ defined in §28, e).

Further, the function $F(x, y, x', y')$ is uniformly continuous in the domain:

$$(x, y) \text{ in } \mathfrak{R}_1 \,, \qquad 1 - a \leqq \sqrt{x'^2 + y'^2} \leqq 1 + a \,,$$

where a is any positive quantity less than 1.

Hence if a positive quantity σ be assigned arbitrarily, the quantities ϵ, ρ and δ can be chosen so small that

$$|F(x_\nu + \tilde{\xi}_\nu, y_\nu + \tilde{\eta}_\nu, \cos\gamma_{\nu+1} + \tilde{\zeta}_\nu, \sin\gamma_{\nu+1} + \tilde{\theta}_\nu) - F(x_\nu, y_\nu, \cos\gamma_{\nu+1}, \sin\gamma_{\nu+1})| < \sigma \,,$$

for $\nu = 0, 1, \cdots, n-1$, and therefore

$$|U_{\text{II}} - W_{\text{II}}| < \sigma \sum_{\nu=0}^{n-1} l_{\nu+1} \,.$$

But if, as we suppose, the curve $\mathfrak{L}$ has a finite length l, we have

$$\sum_{\nu=0}^{n-1} l_{\nu+1} \leqq l \,,$$

and therefore

$$|U_{\text{II}} - W_{\text{II}}| < \sigma l$$

which proves the above statement.

[1] Without multiple points.

$$F_1(x, y, \cos\gamma, \sin\gamma) > 0 \tag{IIa'}$$

is fulfilled.

Then we can construct, according to §28, *d*) and §34, about the extremal $\mathfrak{E}_0$ a field $\mathfrak{S}_k$ which lies in the interior of $\mathfrak{R}$; and if we take k sufficiently small the inequality (IIa′) will be satisfied throughout the region $\mathfrak{S}_k$.

Now let $\mathfrak{L}$ *be any curve of class* (K), *not coinciding with* $\mathfrak{E}_0$, *beginning at A and ending at B, and lying entirely in the interior of* $\mathfrak{S}_k$; let it be represented by (74). *We propose to prove that*

$$J_{\mathfrak{E}_0} < J^*_{\mathfrak{L}}, \tag{81}$$

$J^*_{\mathfrak{L}}$ being defined as in *b*).

Proof:[1] We may apply to the curve $\mathfrak{L}$ the results of *d*), the field $\mathfrak{S}_k$ taking the place of the region there denoted by $\mathfrak{R}_0$.

Accordingly we can choose a partition Π of the interval $(t_0 t_1)$, whose points of division P_ν do not all lie on $\mathfrak{E}_0$, so that the distance

$$|P_\nu P_{\nu+1}| < \rho/3, \qquad (\nu = 0, 1, \cdots, n-1),$$

and that at the same time the arc $P_\nu P_{\nu+1}$ of $\mathfrak{L}$ lies entirely in the circle $(P_\nu, \rho/3)$, where ρ has the same signification as in *d*), and is, moreover, chosen so small that the circle (P_ν, ρ) lies entirely in the interior of $\mathfrak{S}_k$.

We may then, on the one hand, inscribe in $\mathfrak{L}$ a polygon of minimizing extremals with the vertices $A, P_1, P_2, \cdots, P_{n-1}, B$. This polygon is an ordinary curve; it lies entirely in the interior of $\mathfrak{S}_k$, and it does not coincide with $\mathfrak{E}_0$. Hence we have, according to §28, *d*),

$$U_\Pi > J_{\mathfrak{E}_0},$$

say

$$U_\Pi - J_{\mathfrak{E}_0} = p > 0. \tag{82}$$

On the other hand, let Π' be a partition derived from Π by subdivision of the intervals, and so chosen that

$$|U_{\Pi'} - J^*_{\mathfrak{L}}| < p, \tag{83}$$

which is always possible on account of (80). Let $Q_1, Q_2, \cdots, Q_{m-1}$ be the points of division interpolated between the points P_ν and

[1] The outlines of this proof were given by WEIERSTRASS in his *Lectures*, 1879. Another proof has been given by OSGOOD, *Transactions of the American Mathematical Society*, Vol. II (1901), p. 292, by means of the theorem given in §36, *c*).

$P_{\nu+1}$ of the partition Π. These points lie in the circle $(P_\nu, \rho/3)$ and therefore

$$|Q_i Q_{i+1}| \leqq 2\rho/3 \ , \quad (i = 0, 1, \cdots, m-1;\ Q_0 = P_\nu,\ Q_m = P_{\nu+1}) \, .$$

Hence the minimizing extremal from Q_i to Q_{i+1} lies in the circle $(Q_i, 2\rho/3)$ and therefore also in the circle (P_ν, ρ). Hence it follows, according to d), that the minimizing extremal from P_ν to $P_{\nu+1}$ furnishes for the integral J a smaller value than the polygon of minimizing extremals $P_\nu Q_1 Q_2 \cdots Q_{m-1} P_{\nu+1}$, or at most the same value.[1] Therefore

$$U_{\Pi'} \geqq U_{\Pi} \, . \tag{84}$$

But from (82), (83) and (84) follows (81), since we may write

$$J^*_{\mathfrak{L}} - J_{\mathfrak{C}_0} = (J^*_{\mathfrak{L}} - U_{\Pi'}) + (U_{\Pi'} - U_{\Pi}) + (U_{\Pi} - J_{\mathfrak{C}_0}) \, .$$

[1] Viz., when the two curves are identical.

CHAPTER V

KNESER'S THEORY

§32. GAUSS'S THEOREMS ON GEODESICS

KNESER has given, in his "*Lehrbuch der Variationsrechnung*" a new theory of the extremum of the integral

$$J = \int_{t_0}^{t_1} F(x, y, x', y',) \, dt \ ,$$

essentially different from Weierstrass's theory and reaching farther in its results, inasmuch as it furnishes sufficient conditions also for the case when one end-point is movable on a given curve.

Kneser's theory is based upon an extension of certain well-known theorems on geodesics, of which we give—by way of introduction—a brief account in this section.

a) Suppose on a surface there is given a curve $\mathfrak{C}_0$ whose points are determined by a parameter v. At a point $M(v)$ of $\mathfrak{C}_0$ we construct the geodesic $\mathfrak{C}$ normal to $\mathfrak{C}_0$ and lay off on $\mathfrak{C}$ an arc $MP = u$.[1] The position of the end-point P is uniquely determined by the two quantities u, v.

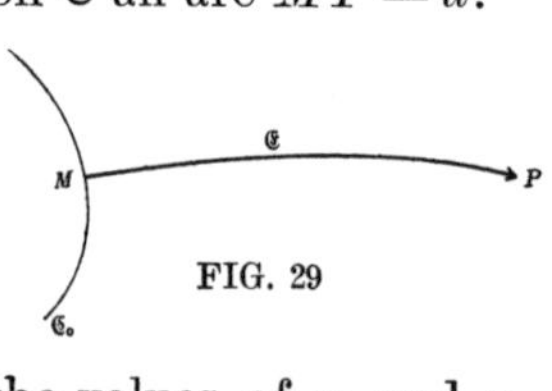

FIG. 29

If we restrict ourselves to such a region $\mathfrak{S}$ of the surface that also conversely P determines uniquely the values of u and v, these two quantities may be introduced as curvilinear co-ordinates on the surface ("geodesic parallel-co-ordinates"). According to a well-known theorem due to GAUSS,[2] *the lines* $u = const.$ *are orthogonal to the geodesics* $v = const.$

[1] *I. e.*, the length of the arc is $|u|$, its direction is determined by the sign of u.

[2] GAUSS, *Disquisitiones generales circa superficies curvas*, art. 16.

b) Hence it follows that the square of the line element takes, for this special system of co-ordinates, the form[1]

$$ds^2 = du^2 + m^2 dv^2 \ .$$

We consider now a particular geodesic, $\mathfrak{E}_0$, of the set $v = \text{const.}$, say $v = v_0$, and on it two points $0 : (u_0, v_0)$ and $1 : (u_1, v_0)$, where $u_0 < u_1$.

We join the points 0 and 1 by an arbitrary curve

$$\overline{\mathfrak{C}}: \qquad \bar{u} = \bar{u}(\tau) \ , \qquad \bar{v} = \bar{v}(\tau) \ , \qquad (\tau_0 \leqq \tau \leqq \tau_1) \ .$$

Then the length of the arc 01 of $\overline{\mathfrak{C}}$ is given by the definite integral

$$\bar{J} = \int_{\tau_0}^{\tau_1} \sqrt{\left(\frac{d\bar{u}}{d\tau}\right)^2 + m^2 \left(\frac{d\bar{v}}{d\tau}\right)^2} \, d\tau \ .$$

On the other hand, the length of the arc 01 of the geodesic $\mathfrak{E}_0$ is

$$J = u_1 - u_0 \ .$$

This may be written

$$J = \int_{\tau_0}^{\tau_1} \frac{d\bar{u}}{d\tau} \, d\tau \ ,$$

and therefore the total variation becomes[2]

$$\Delta J \equiv \bar{J} - J = \int_{\tau_0}^{\tau_1} \left(\sqrt{\left(\frac{d\bar{u}}{d\tau}\right)^2 + m^2 \left(\frac{d\bar{v}}{d\tau}\right)^2} - \frac{d\bar{u}}{d\tau} \right) d\tau \ .$$

The integrand is never negative, and can be zero throughout the whole interval $(\tau_0 \tau_1)$ only when $\overline{\mathfrak{C}}$ coincides with $\mathfrak{E}_0$. Hence it follows that among all curves which can be drawn in $\mathfrak{S}$ between the two points 0 and 1, *the geodesic* $\mathfrak{E}_0$ *has the shortest length.*[3]

It should be noticed that the assumption that the geodesic $\mathfrak{E}_0$ belongs to a set of geodesics satisfying the condi-

[1] GAUSS, *loc. cit.*, art. 19.

[2] Compare DARBOUX, *Théorie des surfaces*, Vol. II, No. 521.

[3] The conclusion can easily be extended to the case where the point 0, instead of being fixed, is movable on a given curve orthogonal to the set of geodesics.

tions imposed upon the region $\mathfrak{S}$, is equivalent to Jacobi's condition.

c) The *necessity* of Jacobi's condition follows from a well-known[1] *theorem on the envelope of a set of geodesics:* If the set of geodesics through the point 0 has an envelope $\mathfrak{F}$, and 02 and 03 are two geodesics of the set touching the envelope at the points 2 and 3, then

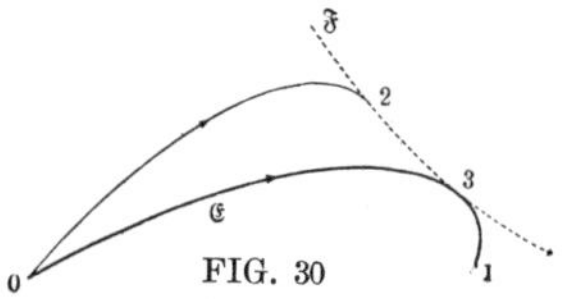

FIG. 30

$$\text{arc } 02 + \text{arc } 23 = \text{arc } 03 \ .$$

The point 3 is the conjugate to 0 on the geodesic 03. Now, if 2 be taken sufficiently near to 3 on the envelope $\mathfrak{F}$, the compound arc 023 is an admissible variation of 03 for which $\Delta J = 0$. And since the envelope $\mathfrak{F}$ is never itself a geodesic,[2] the arc 23 can be replaced by a shorter arc $\overline{23}$, and therefore ΔJ can even be made negative.

Hence the arc 03 does not[3] furnish a minimum, still less an arc 01 of the same geodesic whose end-point 1 lies beyond the conjugate point 3.

The method whose outlines have just been given applies with only slight modifications to the case where only one of the two end-points is given, while the other is movable on a given curve on the surface.

§33. KNESER'S THEOREM ON TRANSVERSALS AND THE THEOREM ON THE ENVELOPE OF A SET OF EXTREMALS

We consider in this section KNESER'S extension to any set of extremals of the two fundamental theorems on sets of geodesics given in the preceding section.

[1] DARBOUX, *Théorie des surfaces*, Vol. II, No. 526, and Vol. III, No. 622.

[2] See DARBOUX, *loc. cit.*, Vol. III, p. 88.

[3] Apart from a certain exceptional case; see §38.

a) Construction of a transversal to a set of extremals: Let

$$x = \phi(t, a) , \qquad y = \psi(t, a) \tag{1}$$

be a set of extremals for the integral

$$J = \int F(x, y, x', y')\, dt ,$$

containing the particular extremal

$$\mathfrak{E}_0: \qquad x = \phi(t, a_0) , \qquad y = \psi(t, a_0) , \qquad t_0 \leqq t \leqq t_1 ,$$

whose minimizing properties are to be investigated. A and B are again the end-points of $\mathfrak{E}_0$.

We suppose that the functions $\phi(t, a)$ and $\psi(t, a)$ are of class C'' in the domain

$$\mathfrak{B}: \qquad T_0 - \epsilon \leqq t \leqq T_1 + \epsilon , \qquad |a - a_0| \leqq d ,$$

where $t_0 - T_0$, $T_1 - t_1$, ϵ and d are positive quantities.

We suppose further that for the extremal $\mathfrak{E}_0$

$$\phi_t^2(t, a_0) + \psi_t^2(t, a_0) \neq 0 \qquad \text{in } (t_0 t_1) . \tag{2}$$

It follows, then, from the continuity of $\phi_t(t, a)$ and $\psi_t(t\ a)$, that the quantities $t_0 - T_0$, $T_1 - t_1$, ϵ, d can be chosen so small that also

$$\phi_t^2(t, a) + \psi_t^2(t, a) \neq 0 \tag{2a}$$

throughout the domain $\mathfrak{B}$.

We denote by $\mathfrak{R}_k$ the rectangle

$$\mathfrak{R}_k: \qquad T_0 \leqq t \leqq T_1 , \qquad |a - a_0| \leqq k < d$$

in the t, a-plane, and by $\mathfrak{S}_k$ its image in the x, y-plane defined by the transformation (1).

To every point (t, a) of $\mathfrak{R}_k$ corresponds a unique point (x, y) of $\mathfrak{S}_k$ which we shall call "the point $[t, a]$." To a continuous curve

$$t = g(\tau) , \qquad a = h(\tau) , \qquad \tau' \leqq \tau \leqq \tau''$$

in $\mathfrak{R}_k$ corresponds a unique curve in $\mathfrak{S}_k$:

$$\tilde{\mathfrak{C}}: \quad \begin{aligned} \tilde{x} &= \phi\big(g(\tau), h(\tau)\big) = \tilde{\phi}(\tau) \ , \\ \tilde{y} &= \psi\big(g(\tau), h(\tau)\big) = \tilde{\psi}(\tau) \ , \end{aligned}$$

which we call[1] the curve $[t=g(\tau),\ a=h(\tau)]$.

The point τ of $\tilde{\mathfrak{C}}$ coincides with the point $t=g(\tau)$ of the extremal $a=h(\tau)$ of the set (1). If for every value of τ the curve $\tilde{\mathfrak{C}}$ is transverse[2] to the extremal $a=h(\tau)$ at their point of intersection, we shall say that $\tilde{\mathfrak{C}}$ is *a transversal to the set of extremals (1)*.

We write for brevity

$$F\big(\phi(t, a)\ ,\quad \psi(t, a)\ ,\quad \phi_t(t, a)\ ,\quad \psi_t(t, a)\big) = \mathbf{F}(t, a)\ , \tag{3}$$

and use the analagous notation for the partial derivatives of F and the function F_1. Then the condition of transversality may be written

$$\mathbf{F}_{x'}(t, a)\frac{d\tilde{x}}{d\tau} + \mathbf{F}_{y'}(t, a)\frac{d\tilde{y}}{d\tau} = 0\ . \tag{4}$$

But

$$\frac{d\tilde{x}}{d\tau} = \phi_t\frac{dt}{d\tau} + \phi_a\frac{da}{d\tau}\ , \qquad \frac{d\tilde{y}}{d\tau} = \psi_t\frac{dt}{d\tau} + \psi_a\frac{da}{d\tau}\ ;$$

hence, remembering the relation (9) of §24, we get

$$\mathbf{F}(t, a)\frac{dt}{d\tau} + \big[\mathbf{F}_{x'}(t, a)\,\phi_a(t, a) + \mathbf{F}_{y'}(t, a)\,\psi_a(t, a)\big]\frac{da}{d\tau} = 0\ . \tag{5}$$

This differential equation for the functions t and a of τ is the necessary and sufficient condition that the curve $\tilde{\mathfrak{C}}$ may be a transversal to the set (1).

We now introduce the further restricting assumption[3] that

$$\mathbf{F}(t, a_0) \neq 0 \qquad \text{in } (t_0 t_1)\ . \tag{6}$$

[1] For the deductions of this section it is not necessary to assume that also conversely to every point (x, y) of $\mathfrak{S}_k$ corresponds a unique point (t, a) of $\mathfrak{R}_k$, provided that we consider the points and curves of $\mathfrak{S}_k$ only in so far as they are the images of definite points and curves of $\mathfrak{R}_k$, and this is what our notation is to indicate. Accordingly two points $[t', a']$ and $[t'', a'']$ of $\mathfrak{S}_k$ are considered as distinct—even if they should have the same co-ordinates x, y—if the points (t', a') and (t'', a'') of $\mathfrak{R}_k$ are distinct.

[2] Compare §30.

[3] We shall free ourselves from this restriction in §37, c).

It follows, then, from the continuity of $\mathbf{F}(t, a)$, that we can take T_0, T_1 so near to t_0, t_1 and k so small that

$$\mathbf{F}(t, a) \neq 0 \tag{6a}$$

throughout the region $\mathfrak{R}_k$.

If the condition (6a) is satisfied, it follows from CAUCHY'S existence theorem[1] on differential equations that *through every point* $[t', a']$ *of the domain* $\mathfrak{S}_k$ *a uniquely*[2] *defined transversal to the set (1) of extremals can be drawn, representable in the form*

$$\left.\begin{aligned} x &= \phi(t, a) \\ y &= \psi(t, a) \end{aligned}\right\} \quad t = \chi(a) \ ,$$

$\chi(a)$ being single-valued and of class C'' in the vicinity of $a = a'$, and taking for $a = a'$ the prescribed value $t = t'$.

The curve $\tilde{\mathfrak{C}}$ may *degenerate*[3] *into a point*, viz., when the functions $\tilde{\phi}(\tau)$, $\tilde{\psi}(\tau)$ reduce to constants, say x^0, y^0. For such a degenerate curve the condition of transversality (4) is evidently always satisfied.

Conversely, if any point (x^0, y^0) in the interior of the region $\mathfrak{R}$ of §24, *b*) is given for which

$$F_1(x^0, y^0, \cos\gamma, \sin\gamma) \neq 0$$

for every γ, and if we construct by the method of §§15 and 27, *c*) the set of extremals through the point (x^0, y^0), this point may always be considered as a degenerate transversal to the set of extremals. For there exists, according to §27, *c*), a function $t^0(a)$ of class C', such that for every a within certain limits

$$x^0 = \phi(t^0, a) \ , \qquad y^0 = \psi(t^0, a) \ ;$$

the point (x^0, y^0) is therefore indeed the image of the curve $t = t^0(a)$ in the t, a-plane.

[1] Compare p. 28, footnote 4.

[2] Compare footnote 1, p. 168.

[3] See KNESER, *Lehrbuch*, p. 47.

b) The function $u(t, a)$*:* Let A^0 be a point on the continuation of $\mathfrak{E}_0$ beyond A, corresponding to an arbitrary value $t = t_0^0$ between T_0 and t_0, and let[1]

$$t = t^0(a)$$

be the transversal $\mathfrak{T}^0$ passing through the point $[t_0^0, a_0]$. We suppose k taken so small that in the interval $(a_0 - k, a_0 + k)$ the function $t^0(a)$ is of class C' and $T_0 < t^0(a) < T_1$. The curve $t = t^0(a)$, interpreted in the t, a-plane, divides the rectangle $\mathfrak{R}_k$ into two regions; we denote by $\mathfrak{R}_k'$ that one for which

$$t \geqq t^0(a) ,$$

and by $\mathfrak{S}_k'$ its image[2] in the x, y-plane.

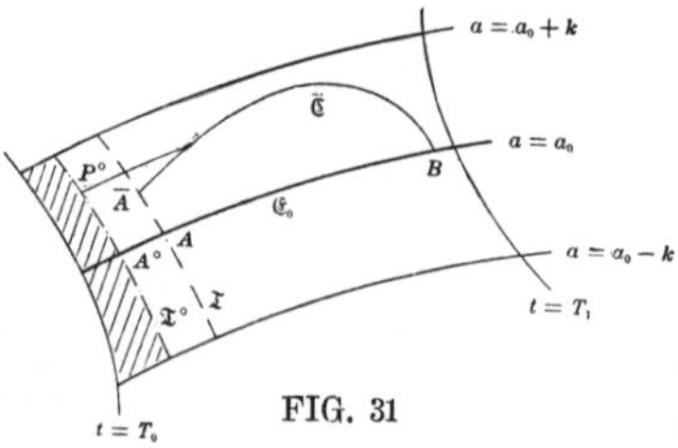

FIG. 31

We consider now any point $P : [t, a]$ of $\mathfrak{S}_k$. The extremal

$$\mathfrak{E}: \quad x = \phi(t, a) , \quad y = \psi(t, a)$$

of the set (1) which passes through P, meets the curve $\mathfrak{T}^0$ at the point $P^0 : [t^0, a]$.

Now denote by u or $u(t, a)$ the value of the definite integral

$$u = \int_{t^0}^{t} \mathbf{F}(t, a)\, dt = u(t, a) \ . \tag{7}$$

The function $u(t, a)$ is single-valued and of class C' in the domain $\mathfrak{R}_k$; moreover it represents,[3] in $\mathfrak{R}_k'$, the value of our integral

$$J = \int F(x, y, x', y')\, dt$$

taken along the extremal $\mathfrak{E}$ from the point P^0 to the point P:

$$u(t, a) = J_{\mathfrak{E}}(P^0 P) \ .$$

[1] When the transversal $\mathfrak{T}^0$ shrinks to a point, the function $t^0(a)$ becomes identical with the function so denoted at the end of a).

[2] In Fig. 31 $\mathfrak{S}_k'$ is the non-shaded part of $\mathfrak{S}_k$.

[3] Only in $\mathfrak{R}_k'$, since we always suppose that the lower limit of the integral J is less than the upper limit; compare §24, b).

The *partial derivatives of* $u(t, a)$ are:

$$\frac{\partial u}{\partial t} = \mathbf{F}(t, a), \tag{8}$$

$$\frac{\partial u}{\partial a} = -\mathbf{F}(t^0, a)\frac{dt^0}{da} + \int_{t^0}^{t} \frac{\partial \mathbf{F}(t, a)}{\partial a}\,dt .$$

But

$$\frac{\partial \mathbf{F}(t, a)}{\partial a} = \mathbf{F}_x \phi_a + \mathbf{F}_y \psi_a + \mathbf{F}_{x'} \phi_{ta} + \mathbf{F}_{y'} \psi_{ta}$$

$$= \frac{\partial}{\partial t}[\mathbf{F}_{x'}\phi_a + \mathbf{F}_{y'}\psi_a] + \phi_a\left[\mathbf{F}_x - \frac{\partial}{\partial t}\mathbf{F}_{x'}\right] + \psi_a\left[\mathbf{F}_y - \frac{\partial}{\partial t}\mathbf{F}_{y'}\right],$$

since $\phi_{ta} = \phi_{at}$, $\psi_{ta} = \psi_{at}$. Now

$$\mathbf{F}_x - \frac{\partial}{\partial t}\mathbf{F}_{x'} = 0 \qquad \text{and} \qquad \mathbf{F}_y - \frac{\partial}{\partial t}\mathbf{F}_{y'} = 0 ,$$

since $\phi(t, a)$ and $\psi(t, a)$ satisfy Euler's differential equation.

Hence we obtain

$$\frac{\partial u}{\partial a} = (\mathbf{F}_{x'}\phi_a + \mathbf{F}_{y'}\psi_a)\Big|^{t} - \left(\mathbf{F}\frac{dt}{da} + \mathbf{F}_{x'}\phi_a + \mathbf{F}_{y'}\psi_a\right)\Big|^{t=t^0} .$$

But the second term disappears since $t = t^0(a)$ represents a transversal and therefore satisfies the differential equation (5).

Thus we finally obtain

$$\frac{\partial u}{\partial a} = \mathbf{F}_{x'}(t, a)\,\phi_a(t, a) + \mathbf{F}_{y'}(t, a)\,\psi_a(t, a) . \tag{9}$$

If the point $P:[t, a]$ moves along a curve $\tilde{\mathfrak{C}}$ defined by[1]

$$t = g(\tau), \qquad a = h(\tau), \qquad i.\,e.,$$

$$\tilde{\mathfrak{C}}: \qquad \begin{cases} \tilde{x} = \phi\big(g(\tau), h(\tau)\big) = \tilde{\phi}(\tau) , \\ \tilde{y} = \psi\big(g(\tau), h(\tau)\big) = \tilde{\psi}(\tau) , \end{cases}$$

u becomes a function of τ whose derivative is, according to (8) and (9):

[1] The functions $g(\tau)$ and $h(\tau)$ are supposed to be of class C' and to furnish points (t, a) in $\mathfrak{R}_k$ so long as τ is restricted to a certain interval $(\tau'\tau'')$ to which we confine ourselves in the following discussion.

$$\frac{du}{d\tau} = \mathbf{F}(t, a)\frac{dt}{d\tau} + \left[\mathbf{F}_{x'}(t, a)\,\phi_a(t, a) + \mathbf{F}_{y'}(t, a)\,\psi_a(t, a)\right]\frac{da}{d\tau},$$

or

$$\frac{du}{d\tau} = \mathbf{F}_{x'}(t, a)\frac{d\tilde{x}}{d\tau} + \mathbf{F}_{y'}(t, a)\frac{d\tilde{y}}{d\tau}\,. \tag{10}$$

The extensions of the two theorems on geodesics of §32 follow immediately from this formula by specializing the curve $\tilde{\mathfrak{E}}$.

c) KNESER'S *Theorem on Transversals:* In the first place we suppose that the curve $\tilde{\mathfrak{E}}$ is a *transversal* to the set (1). Then it follows from (4) and (10) that

$$\frac{du}{d\tau} = 0$$

and therefore $u = \text{const.}$

Thus we obtain the [3]

Theorem I: Two transversals $\mathfrak{T}^0$ and $\mathfrak{T}^1$ to the same set of extremals intercept on the extremals arcs along which the integral J has a constant value.

More explicitly: If $\mathfrak{E}'$ and $\mathfrak{E}''$ are two extremals of the set (1) meeting the transversals $\mathfrak{T}^0$, $\mathfrak{T}^1$ at the points P_0', P_1' and P_0'', P_1'' respectively, then

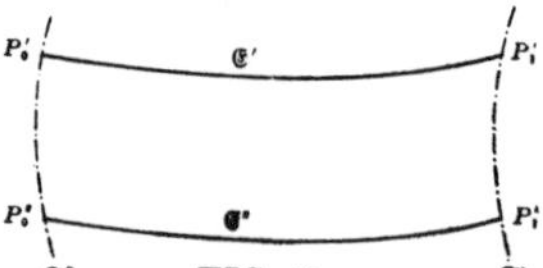

FIG. 32

$$J_{\mathfrak{E}'}(P_0'P_1') = J_{\mathfrak{E}''}(P_0''P_1'')\,. \tag{11}$$

Conversely: If along the curve $\mathfrak{T}^1$ the function $u(t, a)$ is constant, then $\mathfrak{T}^1$ is a transversal of the set (1).

In the special case of the geodesics, transversality is identical with orthogonality,[1] and therefore Kneser's theorem is indeed a generalization of Gauss's theorem on geodesic parallels.

The theorem remains true if one or both of the two transversals shrink to a point;[2] thus we obtain the following corollaries:

[1] Compare §30, *a*).

[2] Compare the remark at the end of *a*).

[3] Compare KNESER, *Lehrbuch*, p. 48.

Corollary I:[1] If $\mathfrak{T}^1$ is a transversal to the set of extremals through a point P_0, then the integral J has the same value if taken along the different extremals from the point P_0 to the curve $\mathfrak{T}^1$, and *vice versa.*

Corollary II: If $\mathfrak{T}^0$ is a transversal to a set of extremals passing through a point P_1, then the integral J has the same value if taken along the different extremals from the curve $\mathfrak{T}^0$ to the point P_1.

Corollary III: If the extremals passing through a point P_0 all pass through a second point P_1, then the integral J has the same value if taken along the different extremals from P_0 to P_1.

d) Theorem on the envelope of a set of extremals: In the second place, we suppose that the curve $\mathfrak{C}$ is *tangent* to all the extremals of the set (1), and therefore is the envelope of the set.

More explicitly: As it has been remarked before, the point τ of $\mathfrak{C}$ coincides with the point $t=g(\tau)$ of the extremal $a=h(\tau)$ of the set (1); we suppose that for every value of τ, at least in a certain interval $(\tau'\tau'')$ in which

$$\left(\frac{d\tilde{x}}{d\tau}\right)^2+\left(\frac{d\tilde{y}}{d\tau}\right)^2 \neq 0 ,$$

the curve $\mathfrak{C}$ and the corresponding extremal are tangent to each other at this common point, so that

$$\begin{vmatrix} \dfrac{d\tilde{x}}{d\tau} & \phi_t \\ \dfrac{d\tilde{y}}{d\tau} & \psi_t \end{vmatrix} = 0 .$$

It follows, then, that there exists a function m of τ such that

$$\phi_t = m\frac{d\tilde{x}}{d\tau} , \qquad \psi_t = m\frac{d\tilde{y}}{d\tau} ;$$

[1] Applied to geodesics, this is GAUSS's theorem on geodesic polar co-ordinates, GAUSS, *loc. cit.*, art. 15.

m is continuous in $(\tau'\tau'')$ and can not change sign.[1] We may without loss of generality[2] suppose that

$$m > 0 \qquad \text{in } (\tau'\tau'') ,$$

i. e., that *the positive directions of the tangents to the two curves coincide.*

From the homogeneity properties of F it follows, then, that

$$\mathbf{F}_{x'}(t, a) = F_{x'}\left(\tilde{x}, \tilde{y}, \frac{d\tilde{x}}{d\tau}, \frac{d\tilde{y}}{d\tau}\right) ,$$

and

$$\mathbf{F}_{y'}(t, a) = F_{y'}\left(\tilde{x}, \tilde{y}, \frac{d\tilde{x}}{d\tau}, \frac{d\tilde{y}}{d\tau}\right) ,$$

and therefore, according to (10),

$$\frac{du}{d\tau} = F\left(\tilde{x}, \tilde{y}, \frac{d\tilde{x}}{d\tau}, \frac{d\tilde{y}}{d\tau}\right) .$$

Hence, integrating from $\tau = \tau'$ to $\tau = \tau''(\tau' < \tau'')$ and remembering the meaning of $u(t, a)$, we obtain the

Theorem II:[3] *Let* $\mathfrak{T}^0$ *be a transversal to the set of extremals (1) and* $\mathfrak{F}$ *the envelope of the set; let, further,* $P'Q'$, $P''Q''$ *be two extremals of the set starting from the points* P', P'' *of* $\mathfrak{T}^0$ *and touching* $\mathfrak{F}$ *at the points* Q', Q'', *then*[4]

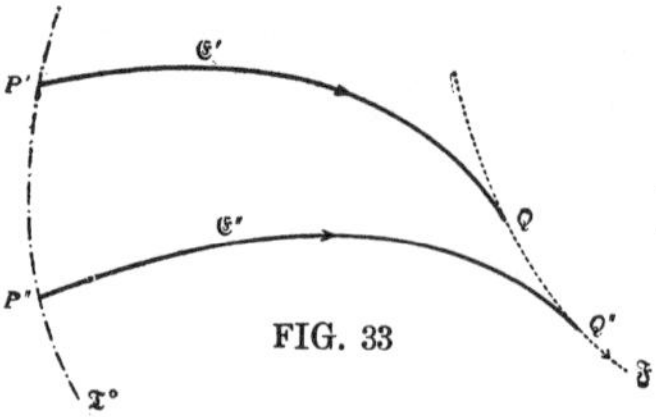

FIG. 33

$$J_{\mathfrak{E}''}(P''Q'') = J_{\mathfrak{E}'}(P'Q') + J_{\mathfrak{F}}(Q'Q'') , \qquad (12)$$

[1] This follows from (2a) and the assumption that

$$\left(\frac{d\tilde{x}}{d\tau}\right)^2 + \left(\frac{d\tilde{y}}{d\tau}\right)^2 \neq 0 \quad \text{in } (\tau'\tau'') .$$

[2] If m is negative, introduce a new parameter

$$\tau = -\sigma \qquad \text{on } \tilde{\mathfrak{E}} .$$

[3] The theorem in the special case when $\mathfrak{T}^0$ shrinks to a point is due to ZERMELO, who proves it by means of Weierstrass's expression for ΔJ in terms of the $\mathbf{E}$-function (*Dissertation*, p. 96). The theorem in its general form and the above proof are due to KNESER; see KNESER, *Lehrbuch*, §25, and also *idem*, *Mathematische Annalen*, Vol. L (1898), p. 27. The simplest case of the theorem is the theorem on the *evolute of a plane curve.*

[4] By a limiting process it can be shown that the theorem remains true if the assumption

with the understanding that the positive direction $Q'Q''$ on $\mathfrak{F}$ has been chosen as indicated above.

The theorem remains true if the transversal $\mathfrak{T}^0$ shrinks to a point, in which case we obtain the *corollary:*

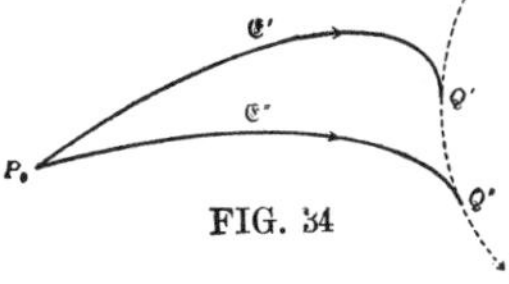

FIG. 34

$$J_{\mathfrak{E}''}(P_0Q'') = J_{\mathfrak{E}'}(P_0Q') + J_{\mathfrak{F}}(Q'Q'') , \quad (13)$$

P_0Q', P_0Q'' being two extremals of the set through P_0, and $\mathfrak{F}$ the envelope of the set.[1]

§34. CONSTRUCTION OF A FIELD

Before we can extend to the general case of extremals the results given in §32, *b*) concerning geodesic parallel co-ordinates, it is necessary to impose upon the set of extremals (1) such further conditions that the correspondence between the two domains $\mathfrak{R}_k$ and $\mathfrak{S}_k$ defined in §33, *a*) becomes a one-to-

$$\left(\frac{d\tilde{x}}{d\tau}\right)^2 + \left(\frac{d\tilde{y}}{d\tau}\right)^2 \neq 0$$

ceases to be satisfied at Q'', *i. e.*, if the curve $\mathfrak{F}$ has a "*cusp*" at Q'', provided that there exists a positive quantity μ such that

$$\frac{d\tilde{x}}{d\tau} \Big/ (\tau''-\tau)^{\mu} \quad \text{and} \quad \frac{d\tilde{y}}{d\tau} \Big/ (\tau''-\tau)^{\mu}$$

approach, for $L\tau = \tau'' - 0$, finite determinate limiting values not both zero (a condition which is, for instance, always fulfilled if $\tilde{x}$ and $\tilde{y}$ are regular in the vicinity of τ''). The proof follows immediately from the homogeneity property of the function F; see §24, (8).

[1] The two theorems on sets of extremals proved in this section can be derived by still a different method indicated for the case of the geodesics by DARBOUX (*Théorie des Surfaces*, Vol. II, No. 536). Let

$$\mathfrak{E}_0: \qquad x = f(t, \alpha_0, \beta_0) , \qquad y = g(t, \alpha_0, \beta_0)$$

be a particular extremal derived from the general solution of Euler's equation, and let $M_0(t=t_0, x=a_0, y=b_0)$ and $M_1(t=t_1, x=a_1, y=b_1)$ be two points on $\mathfrak{E}_0$ which are not conjugate in the more general sense that $\Theta(t_1, t_0) \neq 0$. Then it follows from the theorem on implicit functions that if we take two points $P_0(x_0, y_0)$ and $P_1(x_1, y_1)$ sufficiently near to M_0 and M_1 respectively, a uniquely defined extremal can be drawn through P_0 and P_1:

$$\mathfrak{E}: \qquad x = f(t, \alpha, \beta) , \qquad y = g(t, \alpha, \beta) .$$

The constants α, β, the two values of t which correspond on $\mathfrak{E}$ to the two points

one correspondence, or in other words that the set of extremals (1) furnishes a field about the arc $\mathfrak{C}_0$.

The proof[1] of the existence of a field is based upon the following

Theorem: Let

$$x = \phi(t, a), \qquad y = \psi(t, a) \tag{15}$$

be a one-parameter-set of curves satisfying the following conditions:

A) The functions ϕ and ψ are of class C' in the domain

$$T_0 - \epsilon \leqq t \leqq T_1 + \epsilon, \qquad |a - a_0| \leqq d,$$

ϵ and d being two positive quantities.

B) The particular curve

$$x = \phi(t, a_0), \qquad y = \psi(t, a_0) \tag{16}$$

has no multiple points for $T_0 - \epsilon \leqq t \leqq T_1 + \epsilon$.

C) If we denote by $\Delta(t, a)$ the Jacobian

$$\Delta(t, a) = \frac{\partial(\phi, \psi)}{\partial(t, a)},$$

then

$$\Delta(t, a_0) \neq 0 \qquad \text{in } (T_0 - \epsilon, T_1 + \epsilon). \tag{17}$$

P_0 and P_1, and consequently also the value of the integral J taken from P_0 to P_1 along $\mathfrak{C}$ are single-valued functions of x_0, y_0, x_1, y_1 which are continuous and have continuous partial derivatives in the vicinity of a_0, b_0, a_1, b_1. We denote this integral $J_{\mathfrak{C}}(P_0 P_1)$ considered as a function of x_0, y_0, x_1, y_1, by

$$J(x_0, y_0, x_1, y_1);$$

it is a generalization of the *geodesic distance between two points* (see DARBOUX, *loc. cit.*).

The total differential of this function can be obtained by precisely the same method as that which DARBOUX applies to the geodesic distance, and the result is

$$dJ(x_0, y_0, x_1, y_1) = F_{x'}(x_1, y_1, x_1', y_1')\,dx_1 + F_{y'}(x_1, y_1, x_1', y_1')\,dy_1$$
$$- F_{x'}(x_0, y_0, x_0', y_0')\,dx_0 - F_{y'}(x_0, y_0, x_0', y_0')\,dy_0, \tag{14}$$

the derivatives x_0', y_0' and x_1', y_1' referring to the extremal $\mathfrak{C}$.

Now suppose that P_0 and P_1 move along two curves $\mathfrak{C}_0$ and $\mathfrak{C}_1$ whose co-ordinates are expressed in terms of the same parameter τ. Then the extremals joining corresponding points of $\mathfrak{C}_0$ and $\mathfrak{C}_1$ form a set of extremals with the parameter τ, and $J(x_0, y_0, x_1, y_1)$ changes into a function of τ whose derivative is obtained immediately from (14). By specializing the curves $\mathfrak{C}_0$ and $\mathfrak{C}_1$ the two theorems I and II are obtained.

[1] KNESER'S proof (*Lehrbuch*, §14) must be supplemented by a lemma such as that given below under *a*) and *b*). Compare also OSGOOD, *Transactions of the American Mathematical Society*, Vol. II (1901), p. 277, and BOLZA, *ibid.*, Vol. II (1901), p. 424.

Under these circumstances a positive quantity $k < d$ *can be taken so small that the transformation (15) establishes a one-to-one correspondence between the domain*

$$\mathfrak{R}_k: \qquad T_0 \leqq t \leqq T_1 \ , \qquad |a - a_0| \leqq k$$

in the t, a*-plane, and its image* $\mathfrak{S}_k$ *in the* x, y*-plane.*

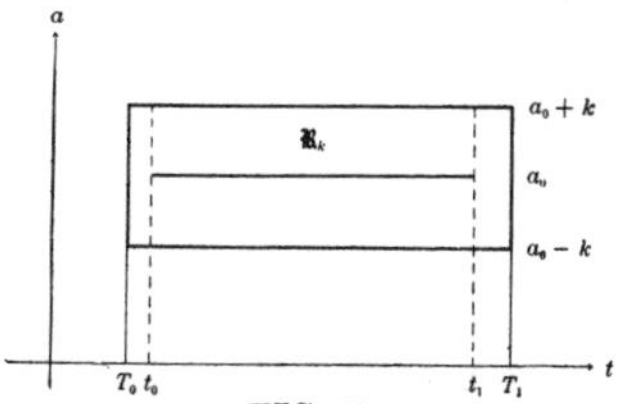

FIG. 35

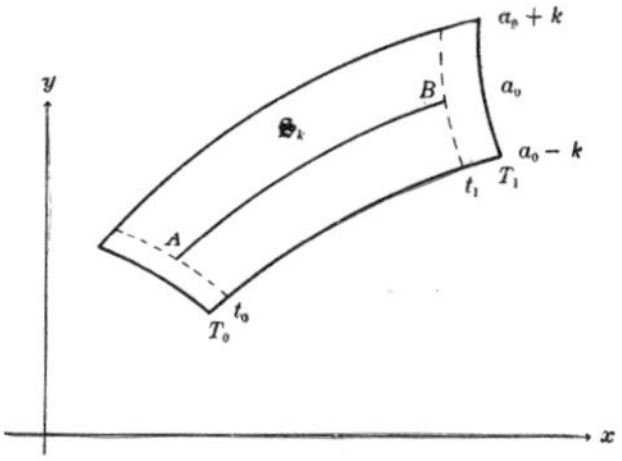

FIG. 36

Proof: We suppose it were not so; that is, we suppose that however small k may be taken, there always exists in $\mathfrak{R}_k$ at least one pair of distinct points (t', a'), (t'', a'') whose images coincide at a point (x, y) of $\mathfrak{S}_k$, and we show that this hypothesis leads to a contradiction to our assumptions.

a) We first select a sequence of decreasing positive quantities

$$k > k_1 > k_2 > \cdots k_\nu > \cdots > 0 \ ,$$

beginning with k and approaching the limit zero, subject to the following rule: After k_1 has been chosen, we select in the rectangle $\mathfrak{R}_{k_1}$ a pair of distinct points $P_1'(t_1', a_1')$ and $P_1''(t_1'', a_1'')$ whose images coincide; this is always possible according to our hypothesis. According to B), a_1' and a_1'' cannot both be equal to a_0; we may therefore choose k_2 smaller than at least one of the two quantities $|a_1' - a_0|$, $|a_1'' - a_0|$, so that at least one of the two points P_1', P_1'' lies outside of $\mathfrak{R}_{k_2}$.

Next we select in $\mathfrak{R}_{k_2}$ a pair of distinct points $P_2'(t_2', a_2')$ and $P_2''(t_2'', a_2'')$ whose images coincide. As before, we can choose k_3 smaller than at least one of the two quantities $|a_2' - a_0|$, $|a_2'' - a_0|$, etc., etc.

Proceeding in this manner, we obtain corresponding to the sequence $\{k_\nu\}$ an infinite sequence of distinct pairs of points

$$P_\nu'(t_\nu', a_\nu') \ , \qquad P_\nu''(t_\nu'', a_\nu'') \ , \qquad \nu = 1, 2, \cdots \infty \ ;$$

the two points P_ν', P_ν'' lie in $\mathfrak{R}_{k_\nu}$, and their images coincide at a point (x_ν, y_ν) of $\mathfrak{S}_k$.

We consider now the set of points

$$\mathfrak{Z} = \{(t_\nu', a_\nu'\,;\ t_\nu'', a_\nu'')\} = \{z_\nu\}$$

in the four-dimensional space $(t', a'\,;\ t'', a'')$. The set $\mathfrak{Z}$ contains an infinitude of distinct points all lying in the finite domain

$$\mathfrak{D}: \qquad T_0 \leqq t' \leqq T_1\,; \ -k \leqq a' - a_0 \leqq k\,;$$
$$T_0 \leqq t'' \leqq T_1\,; \ -k \leqq a'' - a_0 \leqq k\,;$$

it has therefore *at least one accumulation point*[1]

$$\zeta = (\tau', \alpha'\,;\ \tau'', \alpha'') \ ,$$

which belongs itself to $\mathfrak{D}$ since $\mathfrak{D}$ is closed ("abgeschlossen")[3].

b) We are going to prove that

$$\alpha' = a_0 \ , \qquad \alpha'' = a_0 \ , \qquad \tau' = \tau'' \ .$$

Out of the sequence $\{z_\nu\}$ we can select[2] a subsequence $\{z_{\nu_i}\}$ $(i = 1, 2, \cdots \infty\,;\ \nu_{i+1} > \nu_i)$ such that

$$\underset{i=\infty}{L}\, z_{\nu_i} = \zeta \ , \qquad i.\ e.,$$

$$\underset{i=\infty}{L}\, t_{\nu_i}' = \tau' \ , \qquad \underset{i=\infty}{L}\, a_{\nu_i}' = \alpha' \ , \qquad \underset{i=\infty}{L}\, t_{\nu_i}'' = \tau'' \ , \qquad \underset{i=\infty}{L}\, a_{\nu_i}'' = \alpha'' \ .$$

But since $\underset{i=\infty}{L}\, k_{\nu_i} = 0$ and

$$|a_{\nu_i}' - a_0| \leqq k_{\nu_i} \ , \qquad |a_{\nu_i}'' - a_0| \leqq k_{\nu_i} \ ,$$

it follows that

[1] Compare E. I A, p. 185, and II A, p. 45; J. I, No. 27. [2] See J. I, No. 28.
[3] Compare E. I, p. 195.

$$a' = a_0 , \qquad a'' = a_0 ;$$

besides τ' and τ'' are contained in $(T_0 T_1)$.

On the other hand, let $D(t', a'; t'', a'')$ denote the distance between the two points (x', y') and (x'', y'') corresponding to (t', a') and (t'', a''). Then we have

$$D(t'_\nu, a'_\nu ; t''_\nu, a''_\nu) = 0 .$$

But since $D(t', a'; t'', a'')$ is a continuous function of its four arguments, we have

$$D(\tau', a_0 ; \tau'', a_0) = \underset{i=\infty}{L} D(t'_{\nu_i}, a'_{\nu_i} ; t''_{\nu_i}, a''_{\nu_i}) = 0 ,$$

that is, the images (ξ', η') and (ξ'', η'') of the two points (τ', a_0) and (τ'', a_0) coincide. According to B), this is only possible if

$$\tau' = \tau'', \qquad \text{say } = \tau .$$

There exists therefore a point (τ, a_0) in $\mathfrak{R}_k$, in every vicinity of which pairs of distinct points (t', a'), (t'', a'') can be found whose images in the x, y-plane coincide.

c) The theorem on implicit functions[1] leads now immediately to a contradiction. For, let (ξ, η) denote the image of the point (τ, a_0); take (x, y) in the vicinity of (ξ, η) and consider the problem of solving the system of equations

$$x = \phi(t, a) , \qquad y = \psi(t, a)$$

with respect to (t, a). Since $\Delta(\tau, a_0) \neq 0$ it follows from the theorem on implicit functions that after a positive quantity ϵ has been chosen arbitrarily but sufficiently small, a second positive quantity δ_ϵ can be determined such that, if (x, y) be taken in the vicinity (δ_ϵ) of (ξ, η), the above two equations have one and but one solution (t, a) in the vicinity (ϵ) of (τ, a_0).

Further, we can determine, on account of the continuity of ϕ and ψ, a positive quantity $\epsilon' \leqq \epsilon$ such that the image

[1] Compare p. 35, footnote 2.

of every point (t, a) in the vicinity (ϵ') of (τ, a_0) lies in the vicinity (δ_ϵ) of (ξ, η). Hence if (t', a') and (t'', a'') are any two distinct points in the vicinity (ϵ') of (τ, a_0), their images (x', y') and (x'', y'') must lie in the vicinity (δ_ϵ) of (ξ, η) and can therefore not coincide, according to the definition of δ_ϵ.

But this is contrary to the result reached under b); the hypothesis from which we started must therefore be wrong and our theorem is proved.

Corollaries: 1. From the continuity of the functions $\phi(t, a)$, $\psi(t, a)$ and the one-to-one correspondence between $\mathfrak{R}_k$ and $\mathfrak{S}_k$ it follows that the image $\mathfrak{L}'$ of the boundary $\mathfrak{L}$ of the rectangle $\mathfrak{R}_k$ is a continuous closed curve without multiple points (a so-called "*Jordan-curve*"). It divides, therefore,[1] the x, y-plane into an interior and an exterior. According to a theorem due to SCHOENFLIESS[2] *the set of points $\mathfrak{S}_k$ is identical with the interior of $\mathfrak{L}'$ together with the boundary $\mathfrak{L}'$.* Hence it follows that $\mathfrak{S}_k$ is a region in the specific sense of §2, a).

2. Let t_0, t_1 be two values of t satisfying the inequality

$$T_0 < t_0 < t_1 < T_1 \ ,$$

and let $\mathfrak{E}_0$ denote the arc of the curve (16) corresponding to the interval (t_0, t_1). Since the line: $a = a_0$, $t_0 \leqq t \leqq t_1$ lies in the interior of $\mathfrak{R}_k$, its image $\mathfrak{E}_0$ lies in the interior of $\mathfrak{S}_k$ and has, therefore, no point in common with the boundary $\mathfrak{L}'$. The two curves $\mathfrak{E}_0$ and $\mathfrak{L}'$ being continuous, it follows,[3] therefore, that *a neighborhood (ρ) of the arc $\mathfrak{E}_0$ can be constructed which is entirely contained in $\mathfrak{S}_k$.*

3. Since $\Delta(t, a_0) \neq 0$ in $(T_0 T_1)$ and $\Delta(t, a)$ is continuous in $\mathfrak{R}_k$, it follows from the theorem on uniform continuity[4] that *k can be taken so small that*

[1] Compare J. I, No. 102. The interior as well as the exterior is a "continuum."

[2] *Göttinger Nachrichten*, 1899, p. 282; compare also OSGOOD, *ibid.*, 1900, p. 94; and BERNSTEIN, *ibid.*, 1900, p. 98.

[3] Compare p. 13, footnote 4.

[4] Compare E. II A, pp. 18 and 49; P., Nos. 21 and 100; J. I, No. 62.

$$\Delta(t, a) \neq 0 \qquad in \quad \mathfrak{R}_k \, . \tag{18}$$

We suppose in the sequel that k has been selected so small that $\mathfrak{R}_k$ and $\mathfrak{S}_k$ are in a one-to-one correspondence, and that at the same time (18) is satisfied. Under these circumstances the region $\mathfrak{S}_k$ is called *a field about the arc* $\mathfrak{E}_0$, formed by the set of curves (15).

4. The one-to-one correspondence (15) between $\mathfrak{R}_k$ and $\mathfrak{S}_k$ defines t and a as single-valued functions of x and y which are of class C' throughout $\mathfrak{S}_k$; we denote these inverse functions by

$$t = t(x, y) \, , \qquad a = a(x, y) \, . \tag{19}$$

Their derivatives are obtained by the ordinary rules for the differentiation of implicit functions, according to which

$$\begin{aligned} 1 &= \phi_t \frac{\partial t}{\partial x} + \phi_a \frac{\partial a}{\partial x} \, , & 0 &= \phi_t \frac{\partial t}{\partial y} + \phi_a \frac{\partial a}{\partial y} \, , \\ 0 &= \psi_t \frac{\partial t}{\partial x} + \psi_a \frac{\partial a}{\partial x} \, , & 1 &= \psi_t \frac{\partial t}{\partial y} + \psi_a \frac{\partial a}{\partial y} \, . \end{aligned} \tag{20}$$

§35. KNESER'S CURVILINEAR CO-ORDINATES[1]

Our next object is to extend to the general case the results given in §32, *b*) concerning the introduction of geodesic parallel co-ordinates.

a) *Curvilinear co-ordinates in general:* Let us introduce, instead of the rectangular co-ordinates x, y, any system of curvilinear co-ordinates

$$u = U(x, y) \, , \qquad v = V(x, y) \tag{21}$$

where the functions $U(x, y)$ and $V(x, y)$ are of class C'' in a region $\mathfrak{S}$ contained in the region $\mathfrak{R}$ of §24, *b*); in the same region their Jacobian is supposed to be different from zero.

We interpret u, v as the rectangular co-ordinates of a

[1] Compare KNESER, *Lehrbuch*, §16.

point in a u, v-plane and denote by $\mathfrak{T}$ the image in the u, v-plane of the region $\mathfrak{S}$. We suppose, further, that the correspondence established by (21) between $\mathfrak{S}$ and $\mathfrak{T}$ is a one-to-one correspondence. The inverse functions

$$x = X(u, v) \,, \qquad y = Y(u, v) \tag{22}$$

will then likewise be single-valued and of class C'' in the region $\mathfrak{T}$ and moreover their Jacobian

$$D \equiv \frac{\partial(X, Y)}{\partial(u, v)} \neq 0 \qquad \text{in } \mathfrak{T} \,. \tag{23}$$

The image of a region by a transformation of the kind here considered is again a region. Hence $\mathfrak{T}$ and $\mathfrak{S}$ are indeed regions.

We consider now the integral

$$J = \int_{\tau_0}^{\tau_1} F\left(x, y, \frac{dx}{d\tau}, \frac{dy}{d\tau}\right) d\tau \tag{24}$$

taken along an ordinary curve $\mathfrak{C}$: $\quad x = \phi(\tau) \,, \quad y = \psi(\tau)$ from a point $A(\tau_0)$ to a point $B(\tau_1)$, the curve $\mathfrak{C}$ being supposed to lie in the interior of the region $\mathfrak{S}$.

If we introduce the new co-ordinates u, v into the integral J, it will be changed into

$$J' = \int_{\tau_0}^{\tau_1} G\left(u, v, \frac{du}{d\tau}, \frac{dv}{d\tau}\right) d\tau \,, \tag{25}$$

the function G of the four arguments u, v, u', v' being defined by

$$G(u, v, u', v') = F(X, Y, X_u u' + X_v v', Y_u u' + Y_v v') \,. \tag{26}$$

The integral J' is taken along the image $\mathfrak{C}'$ of $\mathfrak{C}$ in the u, v-plane:

$$\mathfrak{C}': \qquad u = U\big(\phi(\tau), \psi(\tau)\big) \,, \qquad v = V\big(\phi(\tau), \psi(\tau)\big)$$

from the point A' (image of A) to the point B' (image of B).

From the equality

$$J' = J \tag{27}$$

it follows that if the curve $\mathfrak{C}$ minimizes[1] the integral J, its

[1] With the understanding that only such curves are admitted as lie in the regions $\mathfrak{S}$ and $\mathfrak{T}$ respectively.

image $\mathfrak{C}'$ necessarily minimizes J', and *vice versa.* Hence the problem to minimize the integral J and the problem to minimize the integral J' may be called *equivalent problems.*

The following properties of the function $G(u, v, u', v')$ can immediately be derived from its definition (26):

1. $G(u, v, u', v')$ is positively homogeneous[1] of dimension 1 in u', v'.

2. By differentiation we get

$$G_{u'} = F_{x'} X_u + F_{y'} Y_u \ ,$$
$$G_{v'} = F_{x'} X_v + F_{y'} Y_v \ .$$

Hence if

$$x' = X_u u' + X_v v' \ , \qquad \grave{x} = X_u \grave{u} + X_v \grave{v} \ ,$$
$$y' = Y_u u' + Y_v v' \ , \qquad \grave{y} = Y_u \grave{u} + Y_v \grave{v} \ ,$$

the following identity holds:

$$\grave{u} G_{u'}(u, v, u', v') + \grave{v} G_{v'}(u, v, u', v')$$
$$= \grave{x} F_{x'}(x, y, x', y') + \grave{y} F_{y'}(x, y, x', y') \ , \tag{28}$$

from which we infer that the **E**-function is an absolute invariant for the transformation (21), *i. e.*, if we denote the new **E**-function by $\mathbf{E}'(u, v;\ u', v';\ \grave{u}, \grave{v})$ we have

$$\mathbf{E}'(u, v;\ u', v';\ \grave{u}, \grave{v}) = \mathbf{E}(x, y;\ x', y';\ \grave{x}, \grave{y}) \ . \tag{29}$$

3. Also F_1 is an invariant; if we denote the corresponding function derived from G by G_1, we obtain easily

$$G_1 = D^2 F_1 \ , \tag{30}$$

where D is defined by (23).

4. Also the left-hand side of Euler's equation is an invariant; after an easy computation, we obtain

$$G_{uv'} - G_{u'v} + G_1(u'v'' - u''v')$$
$$= D\left[F_{xy'} - F_{x'y} + F_1(x'y'' - x''y')\right] \ . \tag{31}$$

The image of an extremal of the old problem is therefore an extremal for the new problem; and the same relation holds for the transversals, as follows from (28).

[1] Compare §24, equation (8).

All these results are in accordance with, and can partly be derived *a priori* from, the equivalence of the two problems.

b) Definition of Kneser's curvilinear co-ordinates: To the assumptions concerning the set of extremals (1) enumerated in §33, *a*), we add the further assumption that

$$\Delta(t, a_0) \neq 0 \qquad \text{in} \quad (t_0 t_1)\ , \tag{32}$$

where $\Delta(t, a)$ denotes again the Jacobian

$$\frac{\partial(\phi, \psi)}{\partial(t, a)}\ .$$

It follows, then, from the continuity of $\Delta(t, a)$, that the quantities $t_0 - T_0$, $T_1 - t_1$, k can be taken so small that

$$\Delta(t, a) \neq 0 \tag{33}$$

throughout the region $\mathfrak{R}_k$.

According to §34, the correspondence between the domains $\mathfrak{R}_k$ and $\mathfrak{S}_k$ defined by (1) is then a one-to-one correspondence, and the inverse functions

$$t = t(x, y)\ , \qquad a = a(x, y) \tag{34}$$

are single-valued and of class C'' in the domain $\mathfrak{S}_k$.

We now combine with the transformation (34) the transformation

$$u = u(t, a)\ , \qquad v = a \tag{35}$$

between the t, a-plane and the u, v-plane, $u(t, a)$ being defined by (7).

Since, according to (6a) and (8),

$$\frac{\partial u}{\partial t} = \mathbf{F}(t, a) \neq 0 \qquad \text{in} \quad \mathfrak{R}_k\ ,$$

it follows that the correspondence between the region $\mathfrak{R}_k$ and its image $\mathfrak{T}_k$ in the u, v-plane, defined by (35), is a one-to-one correspondence and moreover that the Jacobian

$$\frac{\partial(u, v)}{\partial(t, a)} \neq 0 \qquad \text{in} \quad \mathfrak{R}_k\ .$$

Hence, if we combine the two transformations (35) and (34), we obtain a transformation of the form (21) which establishes a one-to-one correspondence between the region $\mathfrak{S}_k$ in the x, y-plane and the region[3] $\mathfrak{T}_k$ in the u, v-plane, and which satisfies all the conditions imposed under a) upon the transformation (21). For every point (x, y) in the region[3] $\mathfrak{S}_k'$ defined in §33, b), the function $u = U(x, y)$ represents, according to the definition of $u(t, a)$ given in §33, the value of the integral J taken along the unique extremal of the set (1) passing through the point (x, y), from the transversal of reference $\mathfrak{T}^0$ to the point (x, y).

c) *Properties of Kneser's curvilinear co-ordinates: For* KNESER'S *curvilinear co-ordinates, the images of the extremals are the lines $v = const.$; the images of the transversals*[1] *the lines $u = const.$ Moreover, the function $G(u, v, u', v')$ has the following characteristic properties:*

$$G(u, v, u', 0) = u' \ ,$$
$$G_{u'}(u, v, u', 0) = 1 \ , \qquad G_{v'}(u, v, u', 0) = 0 \ , \tag{36}$$

which hold for every u, v and for every u' which has the same sign[2] as $\mathbf{F}(t, a)$.

For the proof of these statements it is convenient to represent a curve $\mathfrak{C}$ in the region $\mathfrak{S}_k$ of the x, y-plane in the form

$$\left.\begin{aligned} x &= \phi(t, a) \ , \\ y &= \psi(t, a) \ , \end{aligned}\right\} \begin{aligned} t &= g(\tau) \ , \\ a &= h(\tau) \ , \end{aligned}$$

which is always possible on account of the one-to-one correspondence between $\mathfrak{R}_k$ and $\mathfrak{S}_k$. The image $\mathfrak{C}'$ of $\mathfrak{C}$ in the u, v-plane is then represented by

$$\left.\begin{aligned} u &= u(t, a) \ , \\ v &= a \ , \end{aligned}\right\} \begin{aligned} t &= g(\tau) \ , \\ a &= h(\tau) \ , \end{aligned}$$

and on account of (26) the following identity holds:

[1] Again with the restriction that the transversal must lie in the region $\mathfrak{S}_k'$.

[2] Since $\mathbf{F}(t, a) \neq 0$ and is continuous in $\mathfrak{R}_k$, it has a *constant sign* in $\mathfrak{R}_k$.

[3] Compare p. 182, lines 9–10.

$$F\left(\phi(t,a),\psi(t,a),\frac{d}{d\tau}\phi(t,a),\frac{d}{d\tau}\psi(t,a)\right)$$
$$=G\left(u(t,a),a,\frac{d}{d\tau}u(t,a),\frac{da}{d\tau}\right).$$

If $\mathfrak{C}$ is an extremal of the set (1), it can be defined by the equations

$$t=\tau\ ,\qquad a=a'\ ,$$

a constant.[1] Hence the above formula becomes:

$$\mathbf{F}(\tau,a')=G\left(u(\tau,a'),a',u_\tau(\tau,a'),0\right)\ ,$$

and therefore, on account of (8):

$$u_\tau(\tau,a')=G\left(u(\tau,a'),a',u_\tau(\tau,a'),0\right)\ .$$

Since τ and a' are arbitrary and, moreover,

$$G(u,v,\rho u',0)=\rho G(u,v,u',0)$$

for every positive ρ, the first of the three equations (36) is proved.

The second follows immediately by means of the identity

$$u'G_{u'}+v'G_{v'}=G\ .$$

To prove the third, let

$$t=\tilde{g}(\sigma)\ ,\qquad a=\sigma$$

define a transversal; then, according to §33, c):

$$u\left(\tilde{g}(\sigma),\sigma\right)=\text{const.}$$

Hence the condition of transversality, which must be satisfied at the point of intersection of this transversal with the extremal $t=\tau$, $a=a'$, reduces to

$$\frac{d\tilde{v}}{d\sigma}G_{v'}\left(u(\tau,a'),a',u_\tau(\tau,a'),0\right)=0\ ,$$

from which we infer the third of the equations (36), since

$$\frac{d\tilde{v}}{d\sigma}=1\ .$$

[1] Its image is the line $\mathfrak{C}'$: $u=u(\tau,a')$, $v=a'$ and the angle θ' which the positive direction of $\mathfrak{C}'$ makes with the positive u-axis is 0 or π, according as the constant sign of $\mathbf{F}(t,a)$ is $+$ or $-$.

The relations (36) lead to two important consequences:

In the first place, we obtain immediately from the definition of the $\mathbf{E}'$-function on applying (36):

$$\mathbf{E}'(u, v; u', 0; \grave{u}, \grave{v}) = G(u, v, \grave{u}, \grave{v}) - \grave{u} . \tag{37}$$

In the second place, we get by Taylor's theorem:

$$\begin{aligned} G(u, v, \grave{u}, \grave{v}) - G(u, v, u', 0) \\ = (\grave{u} - u') G_{u'}(u, v, u', 0) + \grave{v} G_{v'}(u, v, u', 0) \\ + \tfrac{1}{2}\left[(\grave{u} - u')^2 \tilde{G}_{u'u'} + 2(\grave{u} - u')\grave{v}\tilde{G}_{u'v'} + \grave{v}^2 \tilde{G}_{v'v}\right] , \end{aligned}$$

where the arguments of $\tilde{G}_{u'u'}$, etc., are

$$u, v, \tilde{u}' = u' + \theta(\grave{u} - u') , \qquad \tilde{v}' = \theta\grave{v} , \qquad \text{and } 0 < \theta < 1 .$$

If we simplify the remainder-term by the introduction of G_1, and make use of (36), we obtain:

$$G(u, v, \grave{u}, \grave{v}) - \grave{u} = \tfrac{1}{2} u'^2 \grave{v}^2 \tilde{G}_1 . \tag{38}$$

From the preceding equation we see that whenever $\tilde{G}_1$ and $\grave{u}$ are both positive (negative), also $G(u, v, \grave{u}, \grave{v})$ is positive (negative). Hence, if for a given point (u, v), the functions $G(u, v, \grave{u}, \grave{v})$ and $G_1(u, v, \grave{u}, \grave{v})$ are different from zero (and therefore do not change sign) for all values of $\grave{u}$, $\grave{v}$ (except possibly $\grave{u} = 0$, $\grave{v} = 0$), they must both have the same sign.

Remembering now the relations (26) and (30), we obtain the following result,[1] which will be useful in the sequel:

If at a point (x, y) the functions $F(x, y, \cos\gamma, \sin\gamma)$ and $F_1(x, y, \cos\gamma, \sin\gamma)$ are both different from zero for all values of γ, then they must both have the same sign.

§ 36. SUFFICIENT CONDITIONS FOR A MINIMUM IN THE CASE OF ONE MOVABLE END-POINT

The introduction of Kneser's curvilinear co-ordinates leads to a number of important consequences:

a) Kneser's sufficient conditions: Through the point A

[1] See KNESER, *Lehrbuch*, p. 53.

(x_0, y_0) of the extremal $\mathfrak{E}_0$ (compare Fig. 31, p. 170) we construct the unique transversal[1] $\mathfrak{T}: [t=\chi(a)]$; and from an arbitrary point $\bar{A}$ of $\mathfrak{T}$ we draw any ordinary curve $\bar{\mathfrak{C}}$, joining the points $\bar{A}$ and B and remaining in the region $\mathfrak{S}'_k$:

$$\bar{\mathfrak{C}}: \qquad \bar{x}=\bar{\phi}(\tau)\,, \qquad \bar{y}=\bar{\psi}(\tau)\,, \qquad \tau_0 \leqq \tau \leqq \tau_1\,.$$

The image of $\mathfrak{E}_0$ in the u, v-plane is the line $v=a_0$; the images of $\mathfrak{T}_0$ and $\mathfrak{T}$ are the lines $u=0$ and $u=u_0\equiv U(x_0, y_0)$; the image of the curve $\bar{\mathfrak{C}}$ is an ordinary curve $\bar{\mathfrak{C}}'$:

$$\bar{\mathfrak{C}}'. \qquad \bar{u}=\bar{u}(\tau)\,, \qquad \bar{v}=\bar{v}(\tau)\,; \qquad \tau_0 \leqq \tau \leqq \tau_1\,.$$

The abscissae u_0 and u_1 of the images A' and B' of A and B are

$$u_0=U(x_0, y_0)\,, \quad u_1=U(x_1, y_1)\,;$$

and according to the definition[2] of $U(x, y)$ we have

$$J_{\mathfrak{E}_0}(AB)=u_1-u_0\,.$$

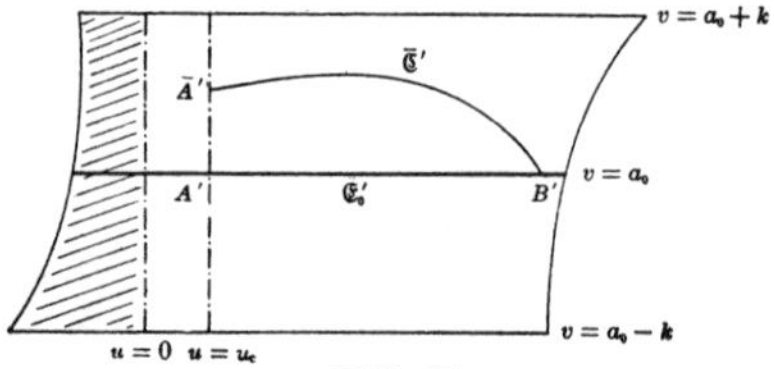

FIG. 37

On the other hand

$$J_{\bar{\mathfrak{C}}}(\bar{A}B)=J'_{\bar{\mathfrak{C}}'}(\bar{A}'B')=\int_{\tau_0}^{\tau_1} G\left(\bar{u}, \bar{v}, \frac{d\bar{u}}{d\tau}, \frac{d\bar{v}}{d\tau}\right) d\tau\,.$$

But since[3] $\bar{u}(\tau_0)=u_0$, $\bar{u}(\tau_1)=u_1$, we have

$$\int_{\tau_0}^{\tau_1} \frac{d\bar{u}}{d\tau}\, d\tau = u_1-u_0\,,$$

and therefore the total variation

$$\Delta J = J_{\bar{\mathfrak{C}}}(\bar{A}B)-J_{\mathfrak{E}_0}(AB)$$

may be written:

$$\Delta J=\int_{\tau_0}^{\tau_1}\left[G\left(\bar{u}, \bar{v}, \frac{d\bar{u}}{d\tau}, \frac{d\bar{v}}{d\tau}\right)-\frac{d\bar{u}}{d\tau}\right] d\tau\,. \tag{39}$$

The relation (38), together with (30), leads now to the following result:

[1] The arc of $\mathfrak{T}$ corresponding to the interval (a_0-k, a_0+k) of a lies entirely in the interior of $\mathfrak{S}'_k$; for A lies in $\mathfrak{S}'_k$ since $t_0 > t_0^0$, and $\mathfrak{T}$ and $\mathfrak{T}^0$ do not intersect in $\mathfrak{S}_k$. The image $\mathfrak{T}'_k$ of $\mathfrak{S}'_k$ is that part of $\mathfrak{T}_k$ in which $u \geqq 0$ or $u \leqq 0$ according as the constant sign of $\mathbf{F}(t, a)$ is + or −.

[2] Compare § 35, b).

[3] Compare, for this important artifice, § 32, b).

If the conditions

$$\Delta(t, a_0) \neq 0 \ , \qquad \mathbf{F}(t, a_0) \neq 0$$

are satisfied for $t_0 \leqq t \leqq t_1$, *and if, moreover,*

$$F_1(x, y, \cos\gamma, \sin\gamma) > 0 \tag{IIa'}$$

along the extremal $\mathfrak{E}_0$ *for every value of* γ, then the extremal $\mathfrak{E}_0$ furnishes for the integral J a smaller value than every other ordinary curve which can be drawn in $\mathfrak{S}_k'$ from the transversal $\mathfrak{T}$ to the point B, provided that k be taken sufficiently small; and therefore *the extremal* $\mathfrak{E}_0$ *minimizes*[1] *the integral* J *if the end-point* B *is to remain fixed while the other end-point is movable on the curve* $\mathfrak{T}$.

b) Weierstrass's theorem for the case of one variable end-point: Still another important conclusion can be derived from (39). On account of (37) we obtain from (39)

$$\Delta J = \int_{\tau_0}^{\tau_1} \mathbf{E}'\left(\bar{u}, \bar{v};\ u', 0;\ \frac{d\bar{u}}{d\tau}, \frac{d\bar{v}}{d\tau}\right) d\tau \ ,$$

where u' is any quantity having the same sign as $\mathbf{F}(t, a)$. We may therefore[2] write the last equation:

$$\Delta J = \int_{\tau_0}^{\tau_1} \mathbf{E}'\left(\bar{u}, v;\ \cos\theta', \sin\theta', \frac{d\bar{u}}{d\tau}, \frac{d\bar{v}}{d\tau}\right) d\tau \ , \tag{40}$$

where θ' is the angle defined on p. 186, footnote 1, and whose value is 0 or π. But since the $\mathbf{E}$-function is, according to (29), an absolute invariant for the transformation (21), we obtain, by returning to the original variables x, y, the *extension of Weierstrass's theorem to the case of one movable end-point:*

$$\Delta J = \int_{\tau_0}^{\tau_1} \mathbf{E}\ (\bar{x}, \bar{y};\ x', y';\ \bar{x}', \bar{y}')\, d\tau \ , \tag{41}$$

[1] To make the connection with the problem: To minimize the integral J by a curve joining a *given* curve $\mathfrak{T}$ with the point B, the following remark is necessary: After an extremal $\mathfrak{E}_0$ of class C' has been found which passes through B, is cut transversely by $\mathfrak{T}$ at A, not touched by $\mathfrak{T}$ at A, then it is always possible, according to §23, *f*) and §30, to determine a set of extremals which has the properties assumed in §33 of the set (1) and to which the curve $\mathfrak{T}$ is a transversal. The transversal $\mathfrak{T}$ of the preceding theory will then coincide with the given curve $\mathfrak{T}$.

[2] Compare §28, equation (51).

where $(\bar{x}, \bar{y})$ is a point of the curve $\bar{\mathfrak{C}}$; $\bar{x}', \bar{y}'$ refer to the curve $\bar{\mathfrak{C}}$; x', y' to the unique extremal of the set (1) passing through the point $(\bar{x}, \bar{y})$.

Reasoning now as in §28, d), we infer that in the above enumeration of sufficient conditions *the condition* (IIa') *may be replaced by the milder condition*

$$\mathbf{E}(x, y;\, p, q;\, \tilde{p}, \tilde{q}) > 0 \qquad \text{along } \mathfrak{C}_0\,, \tag{IV'}$$

understood in the same sense as in §28, d).

c) Osgood's theorem concerning a characteristic property of a strong minimum: The introduction of Kneser's curvilinear co-ordinates leads to a theorem due to OSGOOD[1] concerning the character of the minimum of the integral J, in case the stronger condition (IIa′) is satisfied.

If we denote by $\bar{\theta}$ the angle which the positive tangent to $\bar{\mathfrak{C}}'$ at the point $(\bar{u}, \bar{v})$ makes with the positive u-axis, and introduce on $\bar{\mathfrak{C}}'$ instead of the parameter τ the arc s of $\bar{\mathfrak{C}}'$, we may write (40) in the form[2]

$$\Delta J = \int_{s_0}^{s_1} \mathbf{E}'(\bar{u}, \bar{v};\, \cos\theta', \sin\theta';\, \cos\bar{\theta}, \sin\bar{\theta})\, ds\,.$$

Applying the theorem[3] on the connection between the $\mathbf{E}$-function and F_1 to $\mathbf{E}'$ and G_1, we get

$$\begin{aligned} &\mathbf{E}'(\bar{u}, \bar{v};\, \cos\theta', \sin\theta';\, \cos\bar{\theta}, \sin\bar{\theta}) \\ &\qquad = \big(1 - \cos(\bar{\theta} - \theta')\big)\, G_1(\bar{u}, \bar{v}, \cos\theta^*, \sin\theta^*)\,, \end{aligned}$$

where θ^* is some intermediate value between θ' and $\bar{\theta}$.

Since $\theta' = 0$ or π, the first factor on the right is $1 \mp \cos\bar{\theta}$.

But if we suppose that (IIa′) is satisfied, we can always take k so small that

$$\mathbf{F}_1(x, y, \cos\gamma, \sin\gamma) > 0$$

for every x, y in $\mathfrak{S}_k$ and for every γ.

[1] See *Transactions of the American Mathematical Society*, Vol. II (1901), p. 273. For the following proof see BOLZA, *ibid.*, Vol. II (1901), p. 422.

[2] Compare §28, equation (51).

[3] Compare §28, equation (54).

From the relation (30) between F_1 and G_1, and from the continuity of G_1, it follows, then, that a positive quantity m can be assigned such that

$$G_1(u, v, \cos\omega, \sin\omega) \geqq m$$

for every u, v in $\mathfrak{T}_k$ and for every ω. Accordingly we obtain

$$\Delta J \geqq m \int_{s_0}^{s_1} (1 \mp \cos\bar{\theta})\, ds \ ,$$

or, since

$$\cos\bar{\theta} = \frac{d\bar{u}}{ds} \ ,$$

$$\Delta J \geqq m \left[l \mp (u_1 - u_0) \right] \ ,$$

l being the length of the curve $\bar{\mathfrak{E}}'$ from $\bar{A}'$ to B'.

Now suppose that the curve $\bar{\mathfrak{E}}$ in the x, y-plane passes through a point P of the extremal $a = a_0 + h$ of the set (1), where

$$0 < |h| < k \ .$$

$\bar{\mathfrak{E}}'$ will then pass through a point P' whose ordinate is $v = a_0 + h$.

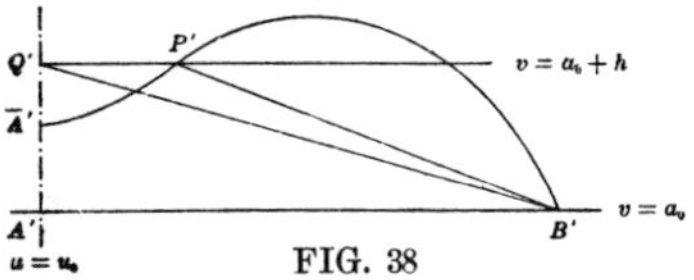

FIG. 38

Let Q' be the foot of the perpendicular from P' upon the line $u = u_0$. Then

$$l \geqq |Q'P'| + |P'B'| \geqq |Q'B'| \ ,$$

that is,

$$l \geqq \sqrt{h^2 + (u_1 - u_0)^2} \ ,$$

and therefore

$$\Delta J \geqq m \left[\sqrt{h^2 + (u_1 - u_0)^2} \mp (u_1 - u_0) \right] > 0 \ . \tag{42}$$

Hence, if we use the symbol $\mathfrak{S}'_h$ in the sense analogous to that of $\mathfrak{S}'_k$, we may formulate the result as follows:

Under our present assumptions concerning the extremal $\mathfrak{E}_0$ and the functions F and F_1, it is always possible to determine, corresponding to every positive quantity h ***numerically*** *less than k, a positive quantity ϵ_h such that*

$$\Delta J = J_{\bar{\mathfrak{E}}}(\bar{A}B) - J_{\mathfrak{E}_0}(AB) \geqq \epsilon_h \tag{43}$$

for every ordinary curve $\bar{\mathfrak{C}}$ which joins the transversal $\mathfrak{T}$ with the point B, and remains within $\mathfrak{S}_k'$ BUT NOT WHOLLY IN THE INTERIOR OF $\mathfrak{S}_h'$.

OSGOOD[1] derives from his theorem a simple proof of WEIERSTRASS's extension[2] of the sufficiency proof to curves without a tangent:

Let, in the notation and terminology of §31, d),

$$\mathfrak{L}: \qquad x = \phi(\tau) \,, \qquad y = \psi(\tau) \,, \qquad \tau_0 \leqq \tau \leqq \tau_1 \,,$$

be a curve of class (K), not coinciding with $\mathfrak{E}_0$, joining the points $\bar{A}$ and B, and lying wholly in the interior of the region $\mathfrak{S}_k'$. Let Π be a partition of the interval $(\tau_0 \tau_1)$ whose subintervals are chosen so small that the corresponding rectilinear polygon $\mathfrak{P}_\Pi$, inscribed in $\mathfrak{L}$, lies in the interior of $\mathfrak{S}_k'$. The polygon being an ordinary curve, we have, if Kneser's sufficient conditions of §36, a) are fulfilled for the extremal $\mathfrak{E}_0$,

$$V_\Pi > J_{\mathfrak{E}_0} \,,$$

if V_Π denotes, as in §31, c), the value of the integral J taken along the polygon $\mathfrak{P}_\Pi$.

Hence if we pass to the limit and remember equation (78) of §31, we obtain

$$J_{\mathfrak{L}}^* \geqq J_{\mathfrak{E}_0} \,.$$

It remains to show that the equality sign cannot take place.

Let Q be any point of $\mathfrak{L}$ not situated on the extremal $\mathfrak{E}_0$, and denote by $a_0 + h$ the value of the parameter a of the extremal of the field passing through Q. Then: $0 < |h| < k$. Now consider in the above limiting process only such partitions Π for which Q is one of the points of division. There exists, then, according to OSGOOD's theorem, a positive quantity ϵ_h such that

$$V_\Pi - J_{\mathfrak{E}_0} \geqq \epsilon_h \,.$$

[1] *Loc. cit.*, p. 292.

[2] Compare § 31, e).

Hence if we pass to the limit,

$$J^*_{\mathfrak{L}} - J_{\mathfrak{E}_0} \geqq \epsilon_h > 0 \ ,$$

and therefore

$$J^*_{\mathfrak{L}} > J_{\mathfrak{E}_0} \ , \qquad \text{Q.E.D.}$$

§37. VARIOUS PROOFS OF WEIERSTRASS'S THEOREM. THE ASSUMPTION $\mathbf{F}(t, a) \neq 0$

The function

$$u = U(x, y)$$

introduced in §35, b) was derived from $u(t, a)$ by substituting for t and a the inverse functions (34):

$$t = t(x, y) \ , \qquad a = a(x, y) \ .$$

Hence *the partial derivatives of* $U(x, y)$ with respect to x and y are, on account of (8) and (9):

$$\frac{\partial U}{\partial x} = \mathbf{F} \frac{\partial t}{\partial x} + (\mathbf{F}_{x'} \phi_a + \mathbf{F}_{y'} \psi_a) \frac{\partial a}{\partial x} \ ,$$

$$\frac{\partial U}{\partial y} = \mathbf{F} \frac{\partial t}{\partial y} + (\mathbf{F}_{x'} \phi_a + \mathbf{F}_{y'} \psi_a) \frac{\partial a}{\partial y} \ .$$

Remembering that

$$\mathbf{F} = \phi_t \mathbf{F}_{x'} + \psi_t \mathbf{F}_{x'}$$

and that by definition

$$\phi\big(t(x, y), a(x, y)\big) \equiv x \ , \qquad \psi\big(t(x, y), a(x, y)\big) \equiv y \ ,$$

we obtain the important *result*:[1]

$$\frac{\partial U}{\partial x} = \mathbf{F}_{x'} = P(x, y) \ ; \qquad \frac{\partial U}{\partial y} = \mathbf{F}_{y'} = Q(x, y) \ , \qquad (44)$$

where $P(x, y)$ and $Q(x, y)$ denote those functions of x and y into which $\mathbf{F}_{x'}(t, a)$ and $\mathbf{F}_{y'}(t, a)$ are transformed when the variables t, a are replaced by their expressions in terms of x, y.

From these expressions of the partial derivatives of U

[1] KNESER, *Lehrbuch*, p. 47; compare also p. 175, footnote 1.

two further proofs of Weierstrass's theorem for the case of one variable end-point, can be derived.

a) Kneser's proof:[1] We repeat the construction of §36, *a*), denoting, however, the points A_0, A, $\bar{A}$, B by numbers: 5, 0, $\bar{0}$, 1 respectively.

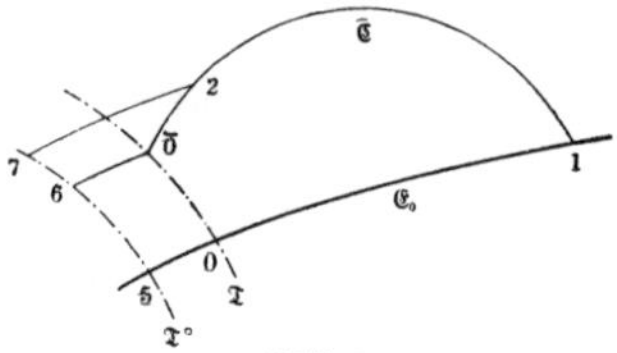

FIG. 39

Then we apply *Weierstrass's construction*[2] slightly modified: Through an arbitrary point $2(\tau=\tau_2)$ of $\bar{\mathfrak{C}}$ we draw the unique extremal of the set (1). It meets the transversal $\mathfrak{T}^0$ at a unique point, 7. Now we consider the integral J taken from 7 along the extremal 72 to 2, and from 2 along the curve $\bar{\mathfrak{C}}$ to 1, and call its value $S(\tau_2)$:

$$S(\tau_2) = J_{72} + \bar{J}_{21} \ ,$$

using the same notation as in §§20 and 28.

In particular we have (see Fig. 39):

$$S(\tau_0) = J_{6\bar{0}} + \bar{J}_{\bar{0}1} \ ,$$
$$S(\tau_1) = J_{50} + J_{01} \ .$$

But according to Kneser's theorem (§33, *c*))

$$J_{6\bar{0}} = J_{50} \ ;$$

hence

$$\Delta J \equiv \bar{J}_{\bar{0}1} - J_{01} = -\left[S(\tau_1) - S(\tau_0)\right] \ .$$

According to the definition of the function $U(x, y)$ given in §35, *b*), we have

$$J_{72} = U(x_2, y_2) \ ;$$

on the other hand

$$\bar{J}_{21} = \int_{\tau_2}^{\tau_1} F(\bar{x}, \bar{y}, \bar{x}', \bar{y}')\, d\tau \ .$$

Hence, making use of (44), we get as in the case of fixed end-points:

[1] KNESER, *Lehrbuch*, §20.

[2] Compare §§20 and 28.

$$\frac{d\,S(\tau_2)}{d\,\tau_2} = -\,\mathbf{E}\,(\overline{x}_2\,,\,\overline{y}_2;\ x_2'\,,\,y_2'\,;\ \overline{x}_2'\,,\,\overline{y}_2')\ . \tag{45}$$

Integrating with respect to τ_2 from τ_0 to τ_1, we obtain *Weierstrass's theorem* (41).

The above deduction leads to the following *geometrical interpretation of the* **E**-*function*, due to KNESER:

Let 3 be the point of $\overline{\mathfrak{C}}$ corresponding to $\tau = \tau_2 + h$, and draw the extremal 83 through the point 3, and the transversal 24 through the point 2 (see Fig. 40). Then

$$S(\tau_2 + h) - S(\tau_2) = J_{84} + J_{43} - J_{72} - \overline{J}_{23}\ ;$$

and since

$$J_{84} = J_{72}\ ,$$

$$S(\tau_2 + h) - S(\tau_2) = J_{43} - \overline{J}_{23}\ .$$

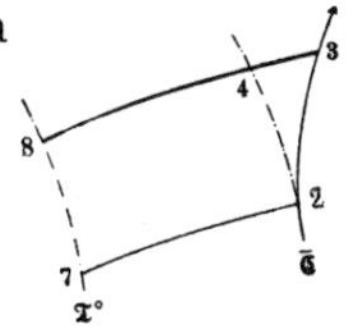

FIG. 40

Hence we obtain, on account of (45), the result:[1]

$$\overline{J}_{23} - J_{43} = h\left[\mathbf{E}\,(\overline{x}_2\,,\,\overline{y}_2\,;\ x_2'\,,\,y_2'\,;\ \overline{x}_2'\,,\,\overline{y}_2') + (h)\right]\ . \tag{46}$$

b) Proof by means of Hilbert's invariant integral: The important formula (44) leads immediately to HILBERT'S invariant integral[2] for the case of parameter-representation.

The integral

$$J^* = \int_{\tau_0}^{\tau_1} \left[P\,(\overline{x}\,,\,\overline{y})\,\overline{x}' + Q\,(\overline{x}\,,\,\overline{y})\,\overline{y}'\right] d\tau\ , \tag{47}$$

taken along $\overline{\mathfrak{C}}$ from $\overline{0}$ to 1 is, according to (44), equal to

$$J^* = \int_{\tau_0}^{\tau_1} \frac{d}{d\tau}\,U\,(\overline{x}\,,\,\overline{y})\,d\tau\ ;$$

hence

$$J^* = U\,(x_1\,,\,y_1) - U\,(\overline{x}_0\,,\,\overline{y}_0)\ ,$$

$\overline{x}_0$, $\overline{y}_0$ denoting the co-ordinates of the point $\overline{0}$.

The value of the integral J^* is therefore *independent of the curve* $\overline{\mathfrak{C}}$ and depends only upon the position of the end-

[1] KNESER, *Lehrbuch*, p. 79; compare footnote 1, p. 138.

[2] Compare §21. Another proof of the invariance of the integral J^*, following more closely the reasoning of HILBERT'S original proof, is given by BLISS, *Transactions of the American Mathematical Society*, Vol. V (1904), p. 121.

points; it even remains *invariant when the point $\bar{0}$ moves along the transversal* $\mathfrak{T}$, since $U(x, y) = \text{const.}$ along every transversal.

Hence, by letting $\bar{0}$ coincide with 0 and $\bar{\mathfrak{C}}$ with $\mathfrak{C}_0$ we obtain

$$J^* = J_{01} .$$

The integral J_{01} can therefore be expressed by an integral taken along the curve $\bar{\mathfrak{C}}$, viz.,

$$J_{01} = \int_{\tau_0}^{\tau_1} \left[F_{x'}(\bar{x}, \bar{y}, x', y')\bar{x}' + F_{y'}(\bar{x}, \bar{y}, x', y')\bar{y}' \right] d\tau .$$

Substituting this value of J_{01} in the difference: $\Delta J = \bar{J}_{\bar{0}1} - J_{01}$ we obtain immediately Weierstrass's theorem.

c) The assumption $\mathbf{F}(t, a) \neq 0$: It is important to notice that in the preceding two proofs of Weierstrass's theorem no use has been made of the assumption (6) that $\mathbf{F}(t, a_0) \neq 0$ at all points of the interval $(t_0 t_1)$, but only of the two special assumptions[1]

$$\mathbf{F}(t_0^0, a_0) \neq 0 , \qquad \mathbf{F}(t_0, a_0) \neq 0 \tag{6b}$$

which, according to §33, *a*), are necessary for the construction of the two transversals $\mathfrak{T}^0$ and $\mathfrak{T}$.

Hence, also in the sufficient conditions derived from Weierstrass's theorem, the condition (6) may be replaced by the milder condition (6b), whereas, in the former deduction of sufficient conditions by means of Kneser's curvilinear co-ordinates, the assumption (6) was essential.

This apparent discrepancy[2] between the two methods can be removed as follows:

[1] The first of these may be replaced by $\mathbf{F}(t, a_0) \not\equiv 0$, because for t_0^0 any value of t between T_0 and t_0 may be chosen. Only in very exceptional cases can F vanish all along an extremal, since the differential equation $F = 0$ is, in general, incompatible with Euler's differential equation.

[2] The discrepancy is still more striking in KNESER's own presentation, since he makes, instead of (6), the stronger assumption

$$F(x, y, \sin\gamma, \cos\gamma) \neq 0$$

along $\mathfrak{C}_0$ for every γ (compare *Lehrbuch*, pp. 49 and 53).

Compare the two problems:

(I) To minimize the integral

$$J = \int_{t_0}^{t_1} F(x, y, x', y')\, dt ,$$

and

(II) To minimize the integral

$$J^{(0)} = \int_{t_0}^{t_1} F^{(0)}(x, y, x', y')\, dt ,$$

where

$$F^{(0)}(x, y, x', y') = F(x, y, x', y') + \Phi_x(x, y)\, x' + \Phi_y(x, y)\, y' , \qquad (48)$$

$\Phi(x, y)$ being a function of x, y alone, of class C' in $\mathfrak{S}_k$. Since

$$J^{(0)} = J + \Phi(x_1, y_1) - \Phi(x_0, y_0) , \qquad (49)$$

we obtain

$$\Delta J^{(0)} = \Delta J$$

for all variations which leave the end-points fixed.

If, on the other hand, the integrals are to be minimized with one end-point, say (x_1, y_1), fixed, while (x_0, y_0) is movable on a given curve $\mathfrak{T}$, the same result holds, provided that $\Phi(x, y)$ remains constant along this curve.

With this condition imposed upon Φ, *the two problems are equivalent;* that is, every solution of the one is also a solution of the other. Hence it follows that every extremal for the one is also an extremal for the other.[1] In particular, our set of curves

$$x = \phi(t, a) , \qquad y = \psi(t, a) \qquad (1)$$

is a set of extremals also for $J^{(0)}$.

We now suppose that the function F satisfies the two conditions (6b), but not (6), and we propose to show that *it is always possible so to select the function* $\Phi(x, y)$ *that*

$$\mathbf{F}^{(0)}(t, a) > 0$$

throughout the region $\mathfrak{R}_k$ defined in §33, a).

[1] The analogous statement for transversals is, in general, not true.

Let m be the minimum of $\mathbf{F}(t, a)$ in the region $\mathfrak{R}_k$, and let M be a positive constant greater than $|m|$.

Further let, as before,

$$t = t(x, y) \;, \qquad a = a(x, y)$$

denote the inverse functions defined in §35, equation (34).

1. *Case of fixed end-points:* In this case we select

$$\Phi(x, y) = M t(x, y) \;. \tag{50}$$

Then

$$\mathbf{F}^{(0)}(t, a) = \mathbf{F}(t, a) + M \frac{\partial}{\partial t} t\big(\phi(t, a), \psi(t, a)\big) \;.$$

But by the definition of the inverse functions we have

$$t\big(\phi(t, a), \psi(t, a)\big) \equiv t \;;$$

hence

$$\mathbf{F}^{(0)}(t, a) = \mathbf{F}(t, a) + M \;,$$

which is positive in $\mathfrak{R}_k$.

2. *Case of one variable end-point:* Suppose (x_1, y_1) fixed and (x_0, y_0) movable along the curve $\mathfrak{T}$, which is a transversal of the set (1) for the problem (I) and represented, as in §36, a), in the form

$$\left.\begin{aligned} x &= \phi(t, a) \;, \\ y &= \psi(t, a) \;, \end{aligned}\right\} t = \chi(a) \;.$$

In this case we select

$$\Phi(x, y) = M\Big[t(x, y) - \chi\big(a(x, y)\big)\Big] \;; \tag{51}$$

then $\Phi(x, y) = 0$ along $\mathfrak{T}$, and

$$\Phi\big(\phi(t, a), \psi(t, a)\big) = M\big(t - \chi(a)\big) \;.$$

Hence we obtain, as before,

$$\mathbf{F}^{(0)}(t, a) = \mathbf{F}(t, a) + M > 0 \qquad \text{in } \mathfrak{R}_k \;.$$

It follows, further, that $\mathfrak{T}$ *is a transversal of the set* (1) *also for problem* (*II*). For

$$\mathbf{F}^{(0)}\frac{dt}{da}+(\phi_a\mathbf{F}_{x'}^{(0)}+\psi_a\mathbf{F}_{y'}^{(0)})$$
$$=\left(\mathbf{F}\frac{dt}{da}+\phi_a\mathbf{F}_{x'}+\psi_a\mathbf{F}_{y'}\right)+\frac{d}{da}\Phi\big(\phi(t,a),\psi(t,a)\big)\ .$$

The first term on the right vanishes for $t=\chi(a)$, since $\mathfrak{T}$ is a transversal of the set (1) for problem (I); the second term vanishes likewise for $t=\chi(a)$, and therefore also the left-hand side, which proves our statement.

The assumption (6), upon which the introduction of Kneser's curvilinear co-ordinates depends, may therefore be made without loss of generality; for, if it should not be satisfied, we can always replace the given problem by an equivalent problem for which it is satisfied.

§38. THE FOCAL POINTS

The assumption $\Delta(t, a_0)\neq 0$ in (t_0t_1) (32)

was indispensable in the previous sufficiency proofs for the construction of a field; but our deductions give no indication whether it is at the same time a necessary condition for a minimum.

We are going to prove, according to KNESER,[1] that at least in the milder form

$$\Delta(t, a_0)\neq 0 \qquad \text{for} \quad t_0<t<t_1\ , \tag{32a}$$

which corresponds to Jacobi's condition in the case of fixed end-points, the condition is indeed necessary for a minimum.

We retain all the assumptions of §33 concerning the set of extremals (1), and we suppose moreover that, in the notation of §33, a),

$$\mathbf{F}_1(t, a_0)>0 \qquad \text{in} \quad (t_0t_1)\ ; \tag{52}$$

but we drop the assumption (32) and suppose, on the contrary, that

[1] KNESER, *Mathematische Annalen*, Vol. L, p. 27, and *Lehrbuch*, §§24, 25.

$$\Delta(t_0', a_0) = 0 \ , \tag{53}$$

where $t_0 < t_0' < t_1$, and, moreover, that t_0' is the smallest value of t, greater than t_0, for which (53) takes place. The corresponding point $A'(x_0', y_0')$ of $\mathfrak{E}_0$ is then the focal point[1] of the transversal $\mathfrak{T}$ on the extremal $\mathfrak{E}_0$.

t_0' is therefore identical with the quantity designated on p. 155 by t_0''. The use of the notation t_0' in the present discussion is justified by the fact that in KNESER's theory the conjugate point appears as a special case of the focal point corresponding to the case when the transversal $\mathfrak{T}$ degenerates into the point A.

a) *Existence of the envelope:* We propose to find all points[2] $[t, a]$ of the x, y-plane in the vicinity of $[t_0', a_0]$ for which

$$\Delta(t, a) = 0 \ . \tag{54}$$

For this purpose we notice in the first place that the function $\Delta(t, a_0)$ is an integral of Jacobi's differential equation

$$F_2 u - \frac{d}{dt}\left(F_1 \frac{du}{dt}\right) = 0 \ .$$

This is proved exactly as the similar statement in §27 *b*) and *c*) by substituting in Euler's differential equation $x = \phi(t, a)$, $y = \psi(t, a)$, differentiating with respect to a and then putting $a = a_0$.

Since $F_1 = \mathbf{F}_1(t, a_0)$ is continuous in the vicinity of $t = t_0'$, and, according to (52), different from zero for $t = t_0'$, it follows that[3]

$$\Delta_t(t_0', a_0) \neq 0 \ . \tag{55}$$

Hence it follows, according to the theorem[4] on implicit functions, that there exists a unique solution $t = \tilde{t}(a)$ of (54) which is of class C′ in the vicinity of $a = a_0$, and takes for $a = a_0$ the value $t = t_0'$.

The curve[5] $[t = \tilde{t}(a)]$ in the x, y-plane, *i. e.*, the curve

[1] Compare §§23 and 30. If $\mathfrak{T}$ shrinks to the point A, the focal point A' becomes the "conjugate" point to A.

[2] For the notation compare §33, *a*).

[3] Compare p. 58, footnote 2.

[4] Compare p. 35, footnote 2.

[5] For the notation, see §33, *a*).

$$\mathfrak{F}: \qquad \tilde{x} = \phi\left(\tilde{t}(a), a\right) = \tilde{\phi}(a) \ , \qquad \tilde{y} = \psi\left(\tilde{t}(a), a\right) = \tilde{\psi}(a)$$

is the *envelope*[1] of the set of extremals (1).

For, since

$$\frac{dx}{da} = \phi_t \frac{d\tilde{t}}{da} + \phi_a \ , \qquad \frac{d\tilde{y}}{da} = \psi_t \frac{d\tilde{t}}{da} + \psi_a \ ,$$

it follows that

$$\frac{d\tilde{x}}{da}\psi_t - \frac{d\tilde{y}}{da}\phi_t = -\Delta\left(\tilde{t}(a), a\right) \equiv 0 \ . \tag{56}$$

This shows, apart from the points at which

$$\left(\frac{d\tilde{x}}{da}\right)^2 + \left(\frac{d\tilde{y}}{da}\right)^2 = 0 \ ,$$

that the curve $\mathfrak{F}$ touches all the extremals of the set (1) for which a is sufficiently near to a_0, and therefore $\mathfrak{F}$ is indeed the envelope of the set.

b) Application of the theorem on envelopes: We must now distinguish two cases:

Case I: The envelope $\mathfrak{F}$ does not degenerate into a point, i. e., $\tilde{\phi}(a)$ and $\tilde{\psi}(a)$ do not both reduce to constants.

Let us suppose that the functions $\tilde{\phi}(a)$ and $\tilde{\psi}(a)$ are of class $C^{(r)}$ in the vicinity of $a = a_0$, that for $a = a_0$ their derivatives up to the order $r-1$ vanish, but that the rth derivatives do not both vanish. Then we obtain by Taylor's formula

$$\frac{d\tilde{x}}{da} = (a - a_0)^{r-1}[A + \alpha] \ , \qquad \frac{d\tilde{y}}{da} = (a - a_0)^{r-1}[B + \beta] \ , \tag{57}$$

where A and B are constants which are not both zero, and α and β approach zero as a approaches a_0.

Substituting these values in (56) we get

$$A = n\phi_t(t_0', a_0) \ , \qquad B = n\psi_t(t_0', a_0) \ , \tag{58}$$

where n is a factor of proportionality which is different from zero.

[1] Compare E., III D, p. 47, footnote 117.

We now introduce on $\mathfrak{F}$ a new parameter τ by the transformation

$$a - a_0 = \epsilon\tau \ ,$$

where $\epsilon = \pm 1$ will be chosen later on. Since, according to (2) and (2a) the functions $\phi_t(\tilde{t}, a)$ and $\psi_t(\tilde{t}, a)$ do not both vanish at $a = a_0$, it follows from (56) that we may write

$$\frac{d\tilde{x}}{d\tau} = m\phi_t \ , \qquad \frac{d\tilde{y}}{d\tau} = m\psi_t \ , \tag{59}$$

where m is a function of τ, which is continuous in the vicinity of $\tau = 0$, and, on account of (57) and (58), is representable in the form

$$m = \epsilon^r \tau^{r-1}(n + \nu) \ ,$$

where $\underset{\tau=0}{L} \nu = 0$.

Whenever it is possible so to select the sign ϵ that m is positive for all sufficiently small negative values of τ, we can construct, according to the theorem II of §33, d), an admissible variation of the arc AA' of $\mathfrak{E}_0$ for which $\Delta J = 0$.

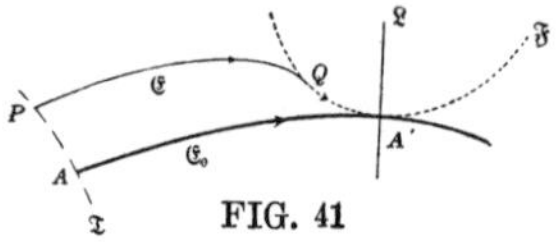

FIG. 41

Subcase A): r odd.[1] If we choose ϵ equal to the sign of n, m is positive for all sufficiently small values of $|\tau|$; see Fig. 41.

Subcase B): r even. m has the same sign as $n\tau$, no matter how we choose ϵ. Therefore

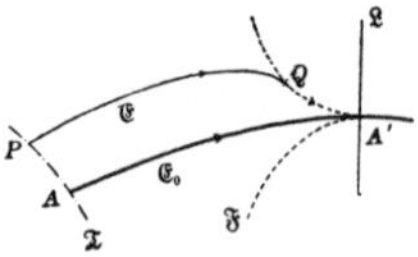

FIG. 42

1. If $n < 0$, m is positive for negative values of τ; see Fig. 42.

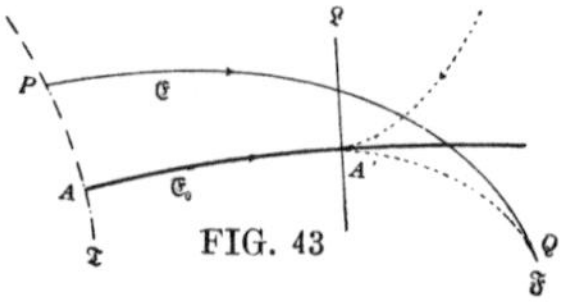

FIG. 43

2. If $n > 0$, m is negative for negative values of τ;[2] see Fig. 43.

In subcase A) and subcase B_1) we have

[1] This covers the "general" case in which $\mathfrak{F}$ has no singular point at $A'(r = 1)$.

[2] If we draw a straight line $\mathfrak{L}$ through the point A' not tangent to $\mathfrak{E}_0$, then $\mathfrak{F}$ crosses the line $\mathfrak{L}$ in case A); it lies all on one side of $\mathfrak{L}$ in case B). on the same side as the arc AA' in case B_1), on the opposite side in case B_2). This follows easily from (57).

$$\Delta J = J_{\mathfrak{E}}(PQ) + J_{\mathfrak{F}}(QA') - J_{\mathfrak{E}_0}(AA') = 0 \ ,$$

according to theorem II of §33, *d*), and therefore the arc AA' of the extremal $\mathfrak{E}_0$ certainly furnishes *no proper*[1] *minimum*, and still less the extremal $\mathfrak{E}_0$ (or AB) itself.

But it furnishes *not even an improper minimum.* For[2] the envelope $\mathfrak{F}$ cannot at the same time be itself an extremal, and therefore the integral $J(QA')$ can be further diminished —and consequently ΔJ can be made negative—by a suitable variation of the arc QA'.

The statement that $\mathfrak{F}$ itself cannot be an extremal can be proved most conclusively by substituting in the left-hand side of Euler's differential equation for x, y the functions

$$\tilde{x} = \phi(\tilde{t}, a) \ , \qquad \tilde{y} = \psi(\tilde{t}, a) \ ,$$

and making use of the characteristic property (59) of the envelope.

If we remember the homogeneity properties of F and its derivatives, and the fact that $\phi(t, a)$, $\psi(t, a)$ as functions of t alone satisfy Euler's differential equation, we obtain after an easy reduction:

$$\tilde{F}_x - \frac{d}{d\tau}\tilde{F}_{x'} = \epsilon \mathbf{F}_1 \Delta_t \psi_t \ ,$$

$$\tilde{F}_y - \frac{d}{d\tau}\tilde{F}_{y'} = \epsilon \mathbf{F}_1 \Delta_t \phi_t \ .$$

The arguments of $\tilde{F}_x$, etc., are

$$\tilde{x}, \tilde{y}, \frac{d\tilde{x}}{d\tau}, \frac{d\tilde{y}}{d\tau} \ ;$$

those of ϕ_t, ψ_t, $\mathbf{F}_1$, Δ_t are $\tilde{t}$, a.

Since, according to our assumptions, $\mathbf{F}_1(t, a)$ and $\Delta_t(t, a)$

[1] For the distinction between "proper" and "improper" minimum, compare §3, *b*).

[2] Compare DARBOUX, *Théorie des Surfaces*, Vol. III, No. 622, and ZERMELO, *Dissertation*, p 96.

are different from zero for $t = t_0'$, $a = a_0$, they remain different from zero in a certain vicinity of this point. Moreover, ϕ_t and ψ_t are not both zero. Hence the envelope $\mathfrak{F}$ does not satisfy Euler's differential equation.[1]

In subcase B_2) the same construction cannot be applied, and therefore the question cannot be decided by this method.

Case II: $\mathfrak{F}$ degenerates into a point. In this case all the extremals of the set pass through the point A', and we can directly apply Corollary II of the theorem on transversals, §33, c).

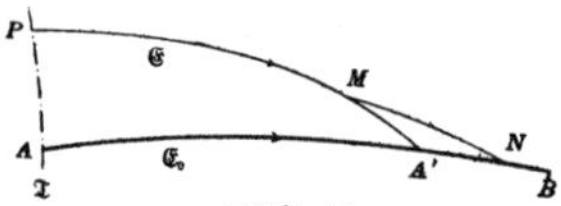

FIG. 44

Accordingly, we have for every extremal $\mathfrak{E}$ of the set:

$$\Delta J = J_{\mathfrak{E}}(PA') - J_{\mathfrak{E}_0}(AA') = 0 ,$$

and therefore the arc AA' of the extremal $\mathfrak{E}_0$ certainly furnishes *no proper minimum.*

Summing up the different cases, we may state the result:

If the end-point B of the extremal AB coincides with the focal point A' (and a fortiori, therefore, if B lies beyond $A' : t_1 > t_0'$) the arc AB ceases to furnish a minimum, except in the following two cases:

1. *When the envelope $\mathfrak{F}$ has at A' a cusp of the special kind defined under subcase B_1)*, the present method fails to give a decision.[2]

2. *When the envelope degenerates into a point*, the arc AA' furnishes no proper minimum, but it may furnish an

[1] Another more geometrical proof can be derived from the fact (see §25, b)) that only one extremal can be drawn through a given point in a given direction if $F_1(x, y, x', y') \neq 0$ for the given point and direction; compare Darboux's proof (*loc. cit.*) for the case of the geodesic.

[2] Under the restricting assumption that $F(x_0', y_0', \cos\gamma, \sin\gamma) \neq 0$ for every γ, OSGOOD has shown that the arc AA' actually furnishes a minimum, if the other sufficient conditions of §36 are satisfied, *Transactions of the American Mathematical Society*, Vol. II (1901), p. 182.

improper minimum.[1] If, however, B lies beyond A', the arc AB furnishes not even an improper minimum.[2]

Thus the necessity of the condition

$$\Delta(t, a_0) \neq 0 \qquad \text{for} \quad t_0 < t < t_1 \tag{32a}$$

is proved for all cases with the one exception just mentioned.[3]

[1] The set of geodesics on a sphere which pass through a point affords an example of this kind.

[2] For, from $\mathbf{F}_1(t_0', a_0) \neq 0$ it follows that if a is sufficiently near to a_0, the "discontinuous solution" $PA'B$ (see Fig. 44) cannot satisfy the corner condition (24) of §25, c) (compare footnote 2, p. 142), and therefore a variation $PMNB$ can be found for which $\Delta J < 0$.

[3] This agrees with the result derived by BLISS from the second variation (compare §30); the latter method proves the necessity of (32a) also in the exceptional case.

CHAPTER VI

ISOPERIMETRIC PROBLEMS[1]

§39. EULER'S RULE

THE special example which has given the name to this class of problems has already been mentioned in §1.

More generally, we understand by an isoperimetric problem one of the following type:

Among all curves joining[2] two given points 0 and 1 for which the definite integral

$$K=\int_{t_0}^{t_1} G(x,y,x',y')\,dt$$

takes a given value l, to determine the one which minimizes (or maximizes) another definite integral

$$J=\int_{t_0}^{t_1} F(x,y,x',y')\,dt\ .$$

Concerning the two functions F and G we make the same assumption as in §24, *b*) concerning F alone. The "admissible curves" are here the totality of ordinary curves which join the two points 0 and 1, lie in the domain $\mathfrak{R}$ of the functions F and G, *and for which the integral K has the given value l.* Aside from this one modification, the definition of a minimum is the same as in the unconditioned problem, §24, *c*). We suppose that a solution has been found:

$$\mathfrak{C}:\qquad x=\phi(t)\ ,\qquad y=\psi(t)\ ,\qquad t_0\leqq t\leqq t_1\ ;$$

and we replace the curve $\mathfrak{C}$ by a neighboring curve

$$\bar{\mathfrak{C}}:\qquad \bar{x}=x+\xi\ ,\qquad \bar{y}=y+\eta\ ,$$

[1] This chapter is based chiefly on WEIERSTRASS's *Lectures* of 1879 and 1882, and on chap. iv of KNESER's book.

[2] Or: joining a given point and a given curve, etc.

where ξ and η are functions of t of class D' satisfying the following conditions:

1. They vanish for $t=t_0$ and $t=t_1$;

2. In the interval (t_0t_1), they remain in absolute value below a certain limit ρ.

3. The integral K taken along $\bar{\mathfrak{C}}$ from t_0 to t_1 has the same value as if taken along $\mathfrak{C}$ (viz., $=l$), or, as we write it,

$$\Delta K = \bar{K}_{01} - K_{01} = 0 \ ; \tag{1}$$

a) Admissible variations: Our next object is to obtain an analytic expression for functions ξ, η satisfying these conditions, not necessarily the most general expression but one of sufficient generality for the purpose of deriving necessary conditions for the minimizing curve.

Such an analytic expression can be obtained, according to WEIERSTRASS, as follows:

Let p_1, p_2, q_1, q_2 be four arbitrary functions of t of class D' vanishing at t_0 and t_1. Then we consider the functions

$$\xi = \epsilon_1 p_1 + \epsilon_2 p_2 \ , \qquad \eta = \epsilon_1 q_1 + \epsilon_2 q_2 \ , \tag{2}$$

where ϵ_1, ϵ_2 are constants, and propose so to determine ϵ_2 as a function of ϵ_1 that the condition (1) is satisfied for every sufficiently small value of ϵ_1.

For this purpose we notice that the integral $\bar{K}_{01}$ is a function of ϵ_1, ϵ_2 which is of class C' in the vicinity of $\epsilon_1=0$, $\epsilon_2=0$, and which is equal to K_{01} for $\epsilon_1=0$, $\epsilon_2=0$. Further, for $\epsilon_1=0$, $\epsilon_2=0$ its partial derivative with respect to ϵ_i has the value

$$N_i = \int_{t_0}^{t_1} (G_x p_i + G_y q_i + G_{x'} p_i' + G_{y'} q_i')\, dt \ .$$

Hence if we introduce the *assumption*[1] *that the curve* $\mathfrak{C}$ *is not an extremal for the integral* K, the functions p_2, q_2 can

[1] If $\mathfrak{C}$ were an extremal for the integral K, the curve $\mathfrak{C}$ (or at least sufficiently small segments of it) would in general minimize or maximize the integral K, and it would therefore be impossible to vary these segments without changing the value of K.

be so chosen that $N_2 \neq 0$, and the conditions of the theorem on implicit functions are fulfilled for the equation (1) in the vicinity of the point $\epsilon_1 = 0$, $\epsilon_2 = 0$. Accordingly, we obtain a unique solution ϵ_2 of the form[1]

$$\epsilon_2 = -\frac{N_1}{N_2}\epsilon_1 + (\epsilon_1)\,\epsilon_1 \,, \tag{3}$$

where (ϵ_1) denotes, as usual, an infinitesimal. Substituting this value in ξ, η we get

$$\begin{aligned} \xi &= \epsilon_1\left(p_1 - \frac{N_1}{N_2}p_2\right) + (\epsilon_1)\,\epsilon_1 p_2 \,, \\ \eta &= \epsilon_1\left(q_1 - \frac{N_1}{N_2}q_2\right) + (\epsilon_1)\,\epsilon_1 q_2 \,. \end{aligned} \tag{4}$$

These functions ξ, η have all the required properties for sufficiently small values of $|\epsilon_1|$. The same argumentation applies to "partial variations" which vary the curve only along a subinterval $(t't'')$ of (t_0t_1). It is only necessary to take the functions p_1, p_2, q_1, q_2 equal to zero in the whole interval (t_0t_1) with the exception of the interior of the subinterval $(t't'')$.

b) Euler's rule: According[2] to §25, the total variation ΔJ for the variations (4) may be written

$$\Delta J = \epsilon_1\left(M_1 - \frac{M_2}{N_2}N_1\right) + (\epsilon_1)\,\epsilon_1 \,,$$

where

$$M_i = \int_{t_0}^{t_1} (F_x p_i + F_y q_i + F_{x'} p_i' + F_{y'} q_i')\, dt \,.$$

For an extremum it is therefore necessary that

$$M_1 - \frac{M_2}{N_2}N_1 = 0 \,.$$

After a definite choice of the functions p_2, q_2 has once been made the quotient M_2/N_2 is a certain numerical constant which we denote by $-\lambda$:

$$\lambda = -\frac{M_2}{N_2} \,. \tag{5}$$

[1] Compare p. 35, footnote 2. [2] Compare, in particular, the footnote on p. 122.

We have then the result that the equation

$$M_1 + \lambda N_1 = 0 \tag{6}$$

must be satisfied for all functions p_1, q_1 of class D' which vanish at t_0 and t_1. This shows at the same time that the value of the constant λ is independent of the choice of the functions p_2, q_2.

If we put

$$H = F + \lambda G \, , \tag{7}$$

equation (6) becomes

$$\int_{t_0}^{t_1} (H_x p_1 + H_y q_1 + H_{x'} p_1' + H_{y'} q_1')\, dt = 0 \; .$$

Hence we infer exactly as in §25 by the method of §6, that *x and y must satisfy the differential equations*

$$H_x - \frac{d}{dt} H_{x'} = 0 \; , \qquad H_y - \frac{d}{dt} H_{y'} = 0 \; , \tag{8}$$

which are equivalent to the one differential equation

$$H_{xy'} - H_{x'y} + H_1(x'y'' - x''y') = 0 \; , \tag{I}$$

where H_1 is defined by:

$$H_1 = \frac{H_{x'x'}}{y'^2} = - \frac{H_{x'y'}}{x'y'} = \frac{H_{y'y'}}{x'^2} \; . \tag{9}$$

We call, again, every curve which satisfies (I) an *extremal* for our problem (KNESER).

The above deduction applies to so-called "discontinuous solutions"[1] as well as to solutions of class C', and shows that *the isoperimetric constant λ has the same constant value along the different segments of a "discontinuous solution."* Moreover we obtain, exactly as in §§9 and 25, at a corner $t = t_2$, the "*corner-condition:*"

[1] Compare §9, in particular footnote 3, p. 37.

[2] This important remark is due to A. MAYER, *Mathematische Annalen*, Vol. XIII (1877), p. 65, footnote; and WEIERSTRASS, *Lectures*. Even if the minimizing curve contains unfrée points or segments, all those segments of the curve whose variation is unrestricted (apart from the condition $\Delta K = 0$) must satisfy the differential equation (I) with the same value of the constant λ.

$$H_{x'}\Big|^{t_2-0} = H_{x'}\Big|^{t_2+0}, \qquad H_{y'}\Big|^{t_2-0} = H_{y'}\Big|^{t_2+0}. \tag{10}$$

All these results may be summarized in the statement that, so far as the first variation is concerned, our problem is equivalent to the problem of minimizing the integral

$$\int_{t_0}^{t_1} (F + \lambda G)\, dt \ ,$$

the curves being subject to no isoperimetric condition.

This simple rule, which is the analogue of a well-known theorem in the theory of ordinary maxima and minima, is usually called *Euler's rule*, according to EULER,[1] who first discovered it.

The rule still holds in the case where the point 0, instead of being fixed, is movable on a given curve

$$\tilde{\mathfrak{C}}: \qquad \tilde{x} = \tilde{\phi}(\tau) \ , \qquad \tilde{y} = \tilde{\psi}(\tau) \ .$$

For, a reasoning similar to that employed in §30, combined with the remark that for all admissible curves

$$\Delta J = \Delta J + \lambda \Delta K \ ,$$

leads[2] to the condition

$$H_{x'}\tilde{x}' + H_{y'}\tilde{y}'\Big|^{t=t_0}_{\tau=\tau_0} = 0 \ . \tag{11}$$

c) EXAMPLE XIII: *Among all curves of given length joining two given points A and B, to determine the one which, together with the chord AB, bounds the maximum area.*

Taking the straight line joining A and B for the x-axis, with BA for positive direction, we have to maximize the integral[3]

$$J = \tfrac{1}{2}\int_{t_0}^{t_1} (xy' - x'y)\, dt$$

[1] EULER, *Methodus inveniendi lineas curvas maximi minimive proprietate gaudentes*, 1744; see STÄCKEL'S translation, p. 101. The first rigorous proof is due to WEIERSTRASS, *Lectures*, and DU BOIS-REYMOND, *Mathematische Annalen*, Vol. XV (1879), p. 310. The proof given in the text is due to WEIERSTRASS.

[2] For details of the proof we refer to KNESER, *Lehrbuch*, §33.

[3] We substitute this analytical problem for the given geometrical one, without entering upon a discussion of the question how far the two are really equivalent. Compare J. I, Nos. 102, 112, and II, Nos. 129–33.

while
$$K = \int_{t_0}^{t_1} \sqrt{x'^2 + y'^2}\, dt$$

has a given value, say l, which we suppose greater than the distance AB.

Since
$$H = \tfrac{1}{2}(xy' - x'y) + \lambda \sqrt{x'^2 + y'^2}\ ,$$

we get
$$H_1 = + \frac{\lambda}{\left(\sqrt{x'^2 + y'^2}\right)^3}\ , \tag{12}$$

and therefore the differential equation (I) becomes
$$\frac{x'y'' - x''y'}{\left(\sqrt{x'^2 + y'^2}\right)^3} = -\frac{1}{\lambda}\ . \tag{13}$$

Hence the radius of curvature of the maximizing curve is constant and has the value $|\lambda|$, while its direction is determined by the sign of λ.

Again, since H_1 never vanishes, there can be no corners,[1] and therefore the curve must be an *arc of a circle of radius* $|\lambda|$. The center and the radius of the circle are determined by the conditions that the arc shall pass through the two given points and shall have the given length l. There are two arcs satisfying these conditions, symmetrical with respect to the x-axis.

d) EXAMPLE XIV: *To draw in a vertical plane between two given points a curve of given length such that its center of gravity shall be as low as possible.*[2]

Taking the positive y-axis vertically upward, we have to minimize the integral
$$J = \int_{t_0}^{t_1} y \sqrt{x'^2 + y'^2}\, dt$$

while at the same time
$$K = \int_{t_0}^{t_1} \sqrt{x'^2 + y'^2}\, dt$$

has a given value, say l.

Here
$$H = (y + \lambda)\sqrt{x'^2 + y'^2}\ .$$

[1] Compare § 25, *c*) and § 28, *b*); in particular footnote 2, p. 142.

[2] Position of equilibrium of a uniform cord suspended at its two extremities.

Using the first of the two differential equations (8), we obtain at once a first integral

$$\frac{x'(y+\lambda)}{\sqrt{x'^2+y'^2}} = c \ .$$

On account of (10), c must have the same constant value all along the curve.

If $c=0$, we obtain[1] the solution

$$x = \text{const.} \ ,$$

which is possible only if the two given points lie in the same vertical line.

If $c \neq 0$, we obtain as general solution of Euler's equation two systems of catenaries:

$$\begin{aligned} x &= \alpha + \beta t \ , \\ y + \lambda &= \pm \beta \cosh t \ . \end{aligned} \tag{14}$$

Determination of the constants. If we suppose $x_0 < x_1$, the constant β must be positive in order that we may have $t_0 < t_1$.

Since the curve is to pass through the two given points, the following equations must be satisfied:

$$\begin{aligned} x_0 &= \alpha + \beta t_0 \ , & y_0 + \lambda &= \pm \beta \cosh t_0 \ , \\ x_1 &= \alpha + \beta t_1 \ , & y_1 + \lambda &= \pm \beta \cosh t_1 \ . \end{aligned}$$

Moreover, the curve must have the given length l; this furnishes the further equation

$$\beta(\sinh t_1 - \sinh t_0) = l \ .$$

From these five equations we have to determine the five constants $\alpha, \beta, \lambda, t_0, t_1$.

If we introduce instead of t_0 and t_1 the two quantities[2]

$$\mu = \frac{t_1 + t_0}{2} = \frac{x_1 + x_0 - 2\alpha}{2\beta} \ ,$$

$$\nu = \frac{t_1 - t_0}{2} = \frac{x_1 - x_0}{2\beta} \ ,$$

we derive from the above equations the following:

[1] $y + \lambda = 0$ is not a solution, since it does not satisfy the second differential equation (8).

[2] WEIERSTRASS, *Lectures*, 1879.

$$
\begin{aligned}
y_1 - y_0 &= \pm\, 2\beta \sinh \mu \sinh \nu \ , \\
l &= \quad 2\beta \cosh \mu \sinh \nu \ .
\end{aligned} \tag{15}
$$

Hence we get

$$\tanh \mu = \pm \frac{y_1 - y_0}{l} \ . \tag{16}$$

Since we suppose

$$l > \sqrt{(x_1 - x_0)^2 + (y_1 - y_0)^2} > |y_1 - y_0| \ ,$$

each of the two equations comprised in (16) has a unique solution μ.

Further, we obtain from (15):

$$l^2 - (y_1 - y_0)^2 = 4\beta^2 \sinh^2 \nu \ ,$$

and therefore

$$\frac{\sinh \nu}{\nu} = \frac{\sqrt{l^2 - (y_1 - y_0)^2}}{x_1 - x_0} \ , \qquad \text{say} = k \ . \tag{17}$$

Since $k > 1$ the transcendental equation (17) has one positive root ν.

After μ and ν have been determined, the values of $\alpha, \beta, \lambda, t_0, t_1$ follow immediately.

Each of the two systems of catenaries (14) contains, therefore, one catenary satisfying the initial conditions.

§40. THE SECOND NECESSARY CONDITION

We suppose that the general solution[1] of the differential equation (I) has been found:

$$x = f(t, \alpha, \beta, \lambda) \ , \qquad y = g(t, \alpha, \beta, \lambda) \ . \tag{18}$$

It contains, besides the two constants of integration α, β, the isoperimetric constant λ.

Moreover, we suppose that a particular system of values of these constants

$$\alpha = \alpha_0 \ , \qquad \beta = \beta_0 \ , \qquad \lambda = \lambda_0$$

has been determined[2] so that the extremal

[1] Compare the remarks in §25, *a*).

[2] There are five equations for the determination of the five unknown quantities $\alpha, \beta, \lambda, t_0, t_1$.

$$\mathfrak{E}_0: \quad \begin{aligned} x &= f(t, \alpha_0, \beta_0, \lambda_0) , \\ y &= g(t, \alpha_0, \beta_0, \lambda_0) , \end{aligned} \qquad t_0 \leqq t \leqq t_1 , \tag{19}$$

passes through the two given points 0 and 1 (for $t = t_0$ and $t = t_1$ respectively), and furnishes for the integral K the prescribed value l:

$$K_{01} = l .$$

We suppose that the functions $f, g, f_t, g_t, f_{tt}, g_{tt}$ and their first partial derivatives with respect to α, β, λ are continuous functions of their four arguments in a domain

$$T_0 \leqq t \leqq T_1 , \qquad |\alpha - \alpha_0| \leqq d , \qquad |\beta - \beta_0| \leqq d , \qquad |\lambda - \lambda_0| \leqq d ,$$

where $T_0 < t_0$ and $T_1 > t_1$.

Further, we assume that for the particular extremal $\mathfrak{E}_0$

$$\begin{aligned} f_t^2 + g_t^2 &\neq 0 \qquad \text{in} \quad (T_0 T_1) , \\ f_t g_\lambda - f_\lambda g_t \Big|^{t_0} &\neq 0 , \end{aligned} \tag{20}$$

and that $f_t g_\alpha - f_\alpha g_t$ and $f_t g_\beta - f_\beta g_t$ are linearly independent.[1]

Finally we retain the assumption introduced in §39 that $\mathfrak{E}_0$ is not an extremal for the integral K.

a) A lemma on a certain type of admissible variations: In §39 the existence of admissible variations of the form

$$\xi = \xi(t, \epsilon) , \qquad \eta = \eta(t, \epsilon) \tag{21}$$

has been established, satisfying the conditions enumerated on p. 122, footnote 1, and besides the isoperimetric condition

$$\Delta K = 0$$

for every sufficiently small value of $|\epsilon|$.

From the latter condition it follows that also

$$\frac{\partial \Delta K}{\partial \epsilon} \equiv 0 .$$

Hence we obtain in particular for $\epsilon = 0$:

[1] Compare §13, end.

$$\int_{t_0}^{t_1} (G_x p + G_y q + G_{x'} p' + G_{y'} q')\, dt = 0 \ , \tag{22}$$

where

$$p = \xi_\epsilon(t, 0) \ , \qquad q = \eta_\epsilon(t, 0) \ . \tag{23}$$

If we transform the left-hand side of (22) by integration by parts, and remember that, as in §25, a),

$$G_x - \frac{d}{dt} G_{x'} \equiv y' U \ , \qquad G_y - \frac{d}{dt} G_{y'} \equiv - x' U \ ,$$

where

$$U \equiv G_{xy'} - G_{x'y} + G_1(x'y'' - x''y') \ ,$$

$$G_1 = \frac{G_{x'x'}}{y'^2} = - \frac{G_{x'y'}}{x'y'} = \frac{G_{y'y'}}{x'^2} \ ,$$

we obtain

$$\int_{t_0}^{t_1} U w \, dt = 0 \ ,$$

where

$$w = y'p - x'q \ .$$

Since p and q vanish at t_0 and t_1, the same is true of w.

Vice versa, the following *lemma*[1] holds:

Let w be any function of class D' which satisfies the conditions

$$w(t_0) = 0 \ , \qquad w(t_1) = 0 \ , \tag{24}$$

$$\int_{t_0}^{t_1} U w \, dt = 0 \ ; \tag{25}$$

then it is always possible to construct an admissible variation of type (21) for which

$$\frac{\partial}{\partial \epsilon}(y'\xi - x'\eta)\Big|^{\epsilon=0} = w \ .$$

Proof: Since $\mathfrak{E}_0$ is not an extremal for the integral K, it follows that $U \not\equiv 0$; it is therefore always possible so to select a function w_1, of class D', and vanishing at t_0 and t_1, that

[1] Due to WEIERSTRASS; see KNESER, *Mathematische Annalen*, Vol. LV, p. 100.

$$\int_{t_0}^{t_1} U w_1 \, dt \neq 0 \ .$$

Now let

$$\omega = \epsilon w + \epsilon_1 w_1 \ ,$$

and choose

$$\xi = \frac{y'\omega}{x'^2 + y'^2} \ , \qquad \eta = \frac{-x'\omega}{x'^2 + y'^2} \ .$$

These functions vanish at t_0 and t_1 for all values of the constants ϵ, ϵ_1; they represent admissible variations if, moreover, the condition

$$\Delta K = 0 \tag{1}$$

is satisfied.

But by the same process as above, we find:

$$\left.\frac{\partial \Delta K}{\partial \epsilon}\right|_{\substack{\epsilon=0\\ \epsilon_1=0}} = \int_{t_0}^{t_1} U w \, dt = 0 \ , \tag{26}$$

$$\left.\frac{\partial \Delta K}{\partial \epsilon_1}\right|_{\substack{\epsilon=0\\ \epsilon_1=0}} = \int_{t_0}^{t_1} U w_1 \, dt \neq 0 \ . \tag{26a}$$

On account of (26a) we can apply the theorem on implicit functions to the equation (1), and obtain for ϵ_1 a unique solution which, on account of (26), is of the form[1]

$$\epsilon_1 = (\epsilon)\,\epsilon \ .$$

Hence

$$y'\xi - x'\eta = \omega = \epsilon w + (\epsilon)\,\epsilon \ ,$$

which proves our statement.

b) Weierstrass's expression for the second variation: Since $\Delta K = 0$, we may write

$$\Delta J = \Delta J + \lambda_0 \Delta K \ . \tag{27}$$

Hence if we apply to the increment $\Delta F + \lambda_0 \Delta G$ Taylor's formula, we obtain for every admissible variation of type (21)

$$\Delta J = \int_{t_0}^{t_1} (H_x \xi + H_y \eta + H_{x'} \xi' + H_{y'} \eta')\, dt + \tfrac{1}{2} \int_{t_0}^{t_1} (H_{xx} \xi^2 + \cdots + H_{y'y'} \eta'^2)\, dt + (\epsilon)\,\epsilon^2 \ ,$$

[1] Compare p. 35, footnote 2.

where $$H = F + \lambda_0 G \ .$$

The first integral is zero since $\mathfrak{E}_0$ is an extremal.

To the second integral we apply the transformation of §27, a). We thus obtain the result:

$$\Delta J = \frac{\epsilon^2}{2}\int_{t_0}^{t_1}\left(H_1\left(\frac{dw}{dt}\right)^2 + H_2 w^2\right)dt + (\epsilon)\,\epsilon^2\ , \tag{28}$$

where H_1 and H_2 are derived from H in the same manner as F_1 and F_2 from F; see §24, b) and §27, a). We shall denote the first term on the right-hand side by $\frac{1}{2}\delta^2 J$.

For a minimum it is therefore necessary that

$$\int_{t_0}^{t_1}\left(H_1\left(\frac{dw}{dt}\right)^2 + H_2 w^2\right)dt \geqq 0\ ; \tag{29}$$

and on account of the lemma proved under a) this condition must be fulfilled *for every function w of class D' which satisfies the equations (24) and (25).*

c) *The second necessary condition:* Since we can construct admissible variations[1] which vary the arc $\mathfrak{E}_0$ only along any given subinterval $(t't'')$ of (t_0t_1), we can apply to the above integral the reasoning of §11, b). Hence *the second necessary condition for a minimum (maximum) is that*

$$H_1 \geqq 0 \qquad (\leqq 0) \tag{II}$$

along the arc $\mathfrak{E}_0$.

This is *the analogue of Legendre's condition.* Also the second necessary condition for the isoperimetric problem coincides, therefore, with the second necessary condition in the problem to minimize the integral

$$\int_{t_0}^{t_1} H(x, y, x', y')\,dt$$

without an isoperimetric condition.

[1] Compare §39, a).

§41. THE THIRD NECESSARY CONDITION AND THE CONJUGATE POINT

We assume in the sequel that (II) is satisfied in the stronger form

$$H_1 > 0 \qquad \text{along } \mathfrak{E}_0 \ . \tag{II$'$}$$

It follows, then, by the method of §11, *b*), that (29) is satisfied, provided that the point 1 is sufficiently near to the point 0.

We have next to determine how near the point 1 must be taken to the point 0 in order that the inequality (29) may remain true. And *it is at this point that the equivalence of the two problems, which we have been comparing, ceases.*[1] In the unconditioned problem the inequality (29) must be fulfilled for all functions w of class D' which vanish at t_0 and t_1; in the isoperimetric problem only for those which besides satisfy the equation (25). It is therefore *a priori* clear that the condition (29) is certainly fulfilled for the isoperimetric problem if it is fulfilled for the unconditioned problem. Hence if we denote by T' the upper limit of the values of t_1 for which the inequality (29) remains true in the isoperimetric problem, by T'' the corresponding upper limit for the unconditioned problem, then T' is at least equal to T'', but it may be greater, and in general it actually is greater, as will be seen later.

a) Determination of the conjugate point: The point T' can be determined by a proper modification, due to WEIERSTRASS, of the method for the determination of the conjugate point in the unconditioned problem:[2] Since we consider only those functions w for which

[1] This has first been discovered by LUNDSTRÖM, "Distinction des maxima et des minima dans un problème isopérimétrique," *Nova acta reg. soc. sc. Upsaliensis*, Ser. 3, Vol. VII (1869); compare also A. MAYER, *Mathematische Annalen*, Vol. XIII (1878), p. 54.

[2] Compare §§12, 13, 16, 27, *b*).

$$\int_{t_0}^{t_1} U w \, dt = 0 \ ,$$

we may write $\delta^2 J$ in the form

$$\delta^2 J = \epsilon^2 \int_{t_0}^{t_1} (H_1 w'^2 + H_2 w^2 + \mu w U) \, dt \ ,$$

μ being an arbitrary constant. Transforming the first term by integration by parts (see §12) and remembering that w vanishes at t_0 and t_1, we obtain, if w' is continuous in $(t_0 t_1)$,

$$\delta^2 J = \epsilon^2 \int_{t_0}^{t_1} w \left[\Psi(w) + \mu U \right] dt \ , \tag{30}$$

where

$$\Psi(w) \equiv H_2 w - \frac{d}{dt}(H_1 w') \ . \tag{31}$$

To obtain the general integral of the differential equation

$$\Psi(w) + \mu U = 0 \tag{32}$$

we substitute in the differential equation[1]

$$H_x - \frac{d}{dt} H_{x'} = 0$$

for x and y the general integral (18), differentiate with respect to a, β, λ respectively, and finally put $a = a_0$, $\beta = \beta_0$, $\lambda = \lambda_0$.

If we denote

$$\left.\begin{aligned} \theta_1(t) &= g_t f_a - f_t g_a \\ \theta_2(t) &= g_t f_\beta - f_t g_\beta \\ \theta_3(t) &= g_t f_\lambda - f_t g_\lambda \end{aligned}\right\} (a = a_0 \ , \quad \beta = \beta_0 \ , \quad \lambda = \lambda_0) \ ,$$

the result[2] is as follows:

[1] H means here: $F + \lambda G$.

[2] For the computation compare §27, b). In the differentiation with respect to λ an additional term appears on account of the factor λ which occurs explicitly in $F + \lambda G$. The immediate result of the differentiation is

$$y' \Psi(\theta_3(t)) + \left(G_x - \frac{d}{dt} G_{x'} \right) = 0 \ ;$$

but according to §25, equation (18),

$$G_x - \frac{a}{dt} G_{x'} = y' U \ ;$$

hence the above result.

$$\Psi\left(\theta_1(t)\right) = 0\ , \quad \Psi\left(\theta_2(t)\right) = 0\ , \quad \Psi\left(\theta_3(t)\right) + U = 0\ . \tag{33}$$

Hence we infer that the function

$$w = c_1\theta_1(t) + c_2\theta_2(t) + \mu\theta_3(t)\ ,$$

in which c_1 and c_2 are arbitrary constants, is the general integral of the differential equation (32).

Now if it were possible to find values for c_1, c_2, μ and a value t' such that

$$w(t_0) \equiv c_1\theta_1(t_0) + c_2\theta_2(t_0) + \mu\theta_3(t_0) = 0\ ,$$
$$w(t') \equiv c_1\theta_1(t') + c_2\theta_2(t') + \mu\theta_3(t') = 0\ ,$$
$$\int_{t_0}^{t'} Uw\,dt \equiv c_1\int_{t_0}^{t'} U\theta_1\,dt + c_2\int_{t_0}^{t'} U\theta_2\,dt + \mu\int_{t_0}^{t'} U\theta_3\,dt = 0\ ,$$
$$t_0 < t' \leqq t_1\ ,$$

the second variation could be made equal to zero (and therefore presumably $\Delta J < 0$) by choosing w equal to zero in $(t't_1)$, and equal to this particular integral in (t_0t').

In order that $\delta^2 J > 0$ for all admissible functions w, it is therefore necessary[1] *that*

$$D(t, t_0) \equiv \begin{vmatrix} \theta_1(t_0) & \theta_2(t_0) & \theta_3(t_0) \\ \theta_1(t) & \theta_2(t) & \theta_3(t) \\ \int_{t_0}^{t} U\theta_1\,dt\ , & \int_{t_0}^{t} U\theta_2\,dt\ , & \int_{t_0}^{t} U\theta_3\,dt \end{vmatrix} \neq 0 \tag{34}$$

for

$$t_0 < t \leqq t_1\ .$$

[1] WEIERSTRASS, *Lectures*, 1872. This condition, together with $H_1 \neq 0$ in (t_0t_1), is also *sufficient for a permanent sign of* $\delta^2 J$ (MAYER, *Mathematische Annalen*, Vol. XIII (1878), p. 53). The proof is based upon the following extension of Jacobi's formula (14) of §12 for the unconditioned problem:

$$(pu+qv)\Psi(pu+qv) = H_1(p'u+q'v)^2 - 2q(p'm+q'n)$$
$$\frac{d}{dt}\left[H_1(pu+qv)(p'u+q'v) - (pm+qn)q\right],$$

where u, v, m, n are the functions introduced below, under b), and p and q are two arbitrary functions. Compare BOLZA, "Proof of the Sufficiency of Jacobi's Condition for a Permanent Sign of the Second Variation in the So-called Isoperimetric Problems," *Transactions of the American Mathematical Society*, Vol. III (1902), p. 305, and *Decennial Publications of the University of Chicago*, Vol. IX, p, 21.

If we denote by t_0' the root next greater than t_0 of the equation[1]

$$D(t, t_0) = 0 \ ,$$

the above inequality (34) may also be written

$$t_1 < t_0' \ .$$

The point t_0' of the extremal $\mathfrak{E}_0$ is again called *the conjugate of the point* t_0.

b) The third necessary condition: The preceding result makes it highly probable[2] that the minimum cannot exist beyond the conjugate point. And indeed it can be proved[3] by a modification of the method employed by WEIERSTRASS for the analogous purpose in the unconditioned problem,[4] that if $t_0' < t_1$, the second variation, and therefore also ΔJ, can be made negative.

For the proof it is convenient to throw the determinant $D(t, t_0)$ into another form in which its properties can be more easily discussed.

Let

$$u \equiv \theta_1(t_0)\,\theta_2(t) - \theta_2(t_0)\,\theta_1(t) \equiv u(t, t_0) \ ,$$
$$v \equiv C_1\theta_1(t) + C_2\theta_2(t) - \theta_3(t) \equiv v(t, t_0) \ ,$$

where the constants C_1, C_2 satisfy the equation

$$C_1\theta_1(t_0) + C_2\theta_2(t_0) - \theta_3(t_0) = 0 \ .$$

These two functions[5] satisfy the two differential equations

[1] $D(t, t_0)$ cannot vanish identically; see below, under *b*).

[2] Compare remarks in §14, p. 59.

[3] The proof has been given by KNESER, *Mathematische Annalen*, Vol. LV (1902), p. 86. From the statements in HORMANN'S *Dissertation* (Göttingen, 1887) it appears that WEIERSTRASS was in possession of essentially the same proof, but I have been unable to ascertain whether he has ever given it in his lectures. I reproduce in the text KNESER'S proof in a slightly simplified form. In §40 of his *Lehrbuch*, KNESER gives another proof which, however, presupposes that $D_t(t_0', t_0) \neq 0$

[4] Compare §16, p. 65, footnote 1.

[5] Neither u nor v can be identically zero. For since, according to (20), $\theta_1(t)$ and $\theta_2(t)$ are linearly independent and $H_1 \neq 0$ in $(t_0 t_1)$, $\theta_1(t_0)$ and $\theta_2(t_0)$ are not both zero, and therefore $u \not\equiv 0$. v cannot be identically zero since $U \not\equiv 0$.

$$\Psi(u) = 0 \ , \qquad \Psi(v) = U \tag{35}$$

respectively, and both vanish at t_0:

$$u(t_0) = 0 \ , \qquad v(t_0) = 0 \ . \tag{36}$$

Hence the determinant $D(t, t_0)$ reduces, after an easy transformation, to

$$D(t, t_0) = mv - nu \ , \tag{37}$$

where

$$m = \int_{t_0}^{t} Uu\,dt \ , \qquad n = \int_{t_0}^{t} Uv\,dt \ .$$

From (35) follows:

$$v\Psi(u) - u\Psi(v) = \frac{d}{dt} H_1(uv' - u'v) = -uU \ .$$

Integrating and remembering (36), we get

$$H_1(uv' - u'v) = -m \ . \tag{38}$$

Again, we obtain by differentiating (37) with respect to t:

$$D' = mv' - nu' \ ,$$

and therefore[1]

$$Du' - D'u = \frac{m^2}{H_1} \tag{39}$$

From the preceding equation it follows that D *has at* t_0' *a zero*[2] *of an odd order, except when* $u(t_0') = 0$.

After these preliminaries, we write the second variation in the form

$$\delta^2 J = -\epsilon^2 k \int_{t_0}^{t_1} w^2\,dt + \epsilon^2 \int_{t_0}^{t_1} w\left[\tilde{\Psi}(w) + \mu U\right] dt \ ,$$

[1] If we denote by t_0'' the root next greater than t_0 of the equation $u(t) = 0$, the relation (39) shows that $t_0' \geqq t_0''$. For, since u has at t_0 a zero only of the first order, the quotient D/u vanishes for t_0, and therefore

$$\frac{D}{u} = -\int_{t_0}^{t} \frac{m^2\,dt}{H_1 u^2} ,$$

which proves that $D \neq 0$ for $t_0 < t < t_0''$.

[2] D cannot vanish identically; otherwise m and therefore also u would vanish identically, which is incompatible with our assumptions.

where k is an arbitrary positive constant and

$$\tilde{\Psi}(w) \equiv (H_2 + k)\, w - \frac{d}{dt}\left(H_1 \frac{dw}{dt}\right) .$$

Now let $\tilde{u}$ and $\tilde{v}$ denote those particular integrals of the differential equations

$$\tilde{\Psi}(\tilde{u}) = 0 \ , \qquad \tilde{\Psi}(\tilde{v}) = U$$

respectively, which satisfy the initial conditions:

$$\tilde{u}(t_0) = u(t_0) = 0 \ , \qquad \tilde{u}'(t_0) = u'(t_0) \ ,$$
$$\tilde{v}(t_0) = v(t_0) = 0 \ , \qquad \tilde{v}'(t_0) = v'(t_0) \ ;$$

then it follows from a general theorem[1] on differential equations containing a parameter that

$$\underset{k=0}{L}\left(\tilde{u}(t) - u(t)\right) = 0 \ , \qquad \underset{k=0}{L}\left(\tilde{v}(t) - v(t)\right) = 0$$

uniformly with respect to the interval $(t_0 t_1)$ *of* t.

Hence, if we put

$$\tilde{m} = \int_{t_0}^{t} U \tilde{u}\, dt \ , \qquad \tilde{n} = \int_{t_0}^{t} U \tilde{v}\, dt \ ,$$
$$\tilde{D}(t, t_0) = \tilde{m}\tilde{v} - \tilde{n}\tilde{u} \ ,$$

we have also

$$\underset{k=0}{L}\, \tilde{D}(t, t_0) = D(t, t_0) \ , \qquad \text{uniformly in } (t_0, t_1) \ .$$

Now suppose that

$$t_0' < t_1$$

and that

$$u(t_0') \neq 0 \ .$$

Then $D(t, t_0)$ changes sign at t_0', as has been shown above; we can therefore choose two quantities t_3 and t_4 satisfying the inequalities

$$t_0 < t_3 < t_0' < t_4 < t_1 \ ,$$

[1] POINCARÉ, *Mécanique céleste*, Vol. I, p. 58; PICARD, *Traité d'Analyse*, Vol. III, p. 167-169; and E. II A, p. 205. The assumption $H_1 \neq 0$ in $(t_0 t_1)$ is essential for this conclusion.

and so near to t_0' that $D(t, t_0)$ has opposite signs at t_3 and t_4. Now select k so small that also $\tilde{D}(t, t_0)$ has opposite signs at t_3 and t_4; then $\tilde{D}(t, t_0)$ vanishes at least once at a point $\tilde{t}_0'$ between t_3 and t_4.

But since $\tilde{D}(\tilde{t}_0', t_0)$ is equal to zero, we can determine two constants c_1, c_2, not both zero, so that

$$c_1 \tilde{u}(\tilde{t}_0') + c_2 \tilde{v}(\tilde{t}_0') = 0 \ ,$$
$$c_1 \tilde{m}(\tilde{t}_0') + c_2 \tilde{n}(\tilde{t}_0') = 0 \ .$$

Now if we choose

$$w = c_1 \tilde{u} + c_2 \tilde{v} \qquad \text{in} \quad (t_0 \tilde{t}_0') \ ,$$
$$w \equiv 0 \qquad \text{in} \quad (\tilde{t}_0' t_1) \ ,$$

and give the arbitrary constant μ the value $-c_2$, then w satisfies the differential equation

$$\tilde{\Psi}(w) + \mu U = 0 \ ,$$

and the conditions (24) and (25).

This function w makes $\delta^2 J$ negative, viz.:

$$\delta^2 J = -\epsilon^2 k \int_{t_0}^{t_1} w^2 \, dt \ .$$

It remains to consider the exceptional case[1] when $u(t_0') = 0$. This can only happen when at the same time $m(t_0') = 0$ and $v(t_0') = 0$, as follows at once from (39) and (38), if we remember that $H_1 \neq 0$ in $(t_0 t_1)$ and that u and u' cannot vanish simultaneously.

In this case we can make $\delta^2 J = 0$ by choosing $\mu = 0$ and

$$w = u \quad \text{in} \ (t_0 t_0') \ , \qquad w \equiv 0 \quad \text{in} \ (t_0' t_1) \ ;$$

and by a slight modification of the method used by SCHWARZ[2] for the proof of the necessity of Jacobi's condition in the unconditioned problem, it can be shown that $\delta^2 J$ can be made negative by choosing

[1] For this exceptional case, see BOLZA, *Mathematische Annalen*, Vol. LVII (1903), p. 44.

[2] Compare §16, p. 65, footnote 1.

$$w = u + ks \quad \text{in } (t_0 t_0') , \qquad w = ks \quad \text{in } (t_0' t_1) ,$$

where

$$s(t_0) = 0 , \qquad s(t_1) = 0 , \qquad s(t_0') \neq 0 ,$$

$$\int_{t_0}^{t_1} s U \, dt = 0 .$$

We thus reach in all cases the result that *the third necessary condition for a minimum is that*

$$D(t, t_0) \neq 0 \qquad \textit{for} \quad t_0 < t < t_1 , \tag{III}$$

or

$$t_0' \geqq t_1 .$$

c) *Kneser's form of the determinant* $D(t, t_0)$: Let $5(t = t_{50})$ be a point on the continuation of the extremal $\mathfrak{E}_0$ beyond the point 0, taken sufficiently near to 0, or else the point 0 itself. Then it follows from our assumptions concerning the general solution (18) of the differential equation (I) that there exists[1] a doubly infinite system Σ of extremals passing through the point 5:

$$x = \phi(t, a, b) , \qquad y = \psi(t, a, b) , \tag{40}$$

and satisfying the following conditions:

1. The extremal $\mathfrak{E}_0$ is contained in the system Σ, say for $a = a_0$, $b = b_0$.

2. The functions

$$\phi, \psi, \phi_t, \psi_t, \phi_{tt}, \psi_{tt}$$

and their first partial derivatives with respect to a and b are continuous in a domain

$$T_0 \leqq t \leqq T_1 , \qquad |a - a_0| \leqq d_1 , \qquad |b - b_0| \leqq d_1 , \tag{41}$$

where $T_0 < t_{50} < t_0 < t_1 < T_1$ and d_1 is a sufficiently small positive constant.

3. $\phi_t^2 + \psi_t^2 \neq 0$ in the domain (41).

4. The value $t = t_5$, to which corresponds on the extremal (a, b)

[1] If $x = f(t, \alpha, \beta, \lambda)$, $y = g(t, \alpha, \beta, \lambda)$ represents an extremal passing through the point 5 (say for $t = t_5$), the quantities $\alpha, \beta, \lambda, t_5$ must satisfy the two equations

$$f(t_5, \alpha, \beta, \lambda) - f(t_{50}, \alpha_0, \beta_0, \lambda_0) = 0 ,$$
$$g(t_5, \alpha, \beta, \lambda) - g(t_{50}, \alpha_0, \beta_0, \lambda_0) = 0 .$$

Solving with respect to t_5 and λ and remembering (20), we obtain the results stated in the text.

the point 5, is a function of a and b, of class C' in the vicinity of a_0, b_0.

From the definition of t_5, according to which,

$$x_5 = \phi(t_5, a, b) , \qquad y_5 = \psi(t_5, a, b) ,$$

it follows by differentiation that

$$\begin{aligned} \phi_t\Big|^{t_5} \frac{\partial t_5}{\partial a} + \phi_a\Big|^{t_5} = 0 , \qquad \psi_t\Big|^{t_5} \frac{\partial t_5}{\partial a} + \psi_a\Big|^{t_5} = 0 , \\ \phi_t\Big|^{t_5} \frac{\partial t_5}{\partial b} + \phi_b\Big|^{t_5} = 0 , \qquad \psi_t\Big|^{t_5} \frac{\partial t_5}{\partial b} + \psi_b\Big|^{t_5} = 0 . \end{aligned} \tag{42}$$

5. λ is a function of a, b of class C' in the vicinity of a_0, b_0, and the two derivatives

$$\lambda_1 = \lambda_a(a_0, b_0) , \qquad \lambda_2 = \lambda_b(a_0, b_0)$$

are not both zero, since $\theta_1(t)$ and $\theta_2(t)$ are two linearly independent integrals of $\Psi(u)=0$ (compare (33)).

We shall denote by

$$\mathbf{F}(t, a, b) , \quad \mathbf{G}(t, a, b) , \quad \mathbf{H}(t, a, b) , \quad \mathbf{G}_x(t, a, b) , \quad \text{etc.}$$

the functions of t, a, b into which F, G, H, G_x, etc., change on substituting

$$x = \phi(t, a, b) , \qquad y = \psi(t, a, b) ,$$
$$x' = \phi_t(t, a, b) , \qquad y' = \psi_t(t, a, b) .$$

The integral K taken along any extremal (a, b) of the system Σ from the point $5(t=t_5)$ to an arbitrary point t, is a function of t, a, b, which we denote by $\chi(t, a, b)$:

$$\chi(t, a, b) = \int_{t_5}^{t} \mathbf{G}(t, a, b)\, dt . \tag{43}$$

Finally we denote by $\Delta(t, a, b)$ the Jacobian of ϕ, ψ, χ:

$$\Delta(t, a, b) = \frac{\partial(\phi, \psi, \chi)}{\partial(t, a, b)},$$

Then Weierstrass's function $D(t, t_{50})$ *differs from the Jacobian* $\Delta(t, a_0, b_0)$ *only by a constant factor:*

$$D(t, t_{50}) = C\Delta(t, a_0, b_0) . \tag{44}$$

Proof: For the partial derivatives of $\chi(t, a, b)$ we obtain the following values

$$\chi_t = \mathsf{G} = \phi_t \mathsf{G}_{x'} + \psi_t \mathsf{G}_{y'} \ ,$$

$$\chi_a = \int_{t_5}^{t} (\mathsf{G}_x \phi_a + \mathsf{G}_y \psi_a + \mathsf{G}_{x'} \phi_{ta} + \mathsf{G}_{y'} \psi_{ta})\, dt - \mathsf{G}\Big|^{t_5} \frac{\partial t_5}{\partial a} \ .$$

Applying the usual integration by parts and remembering that[1]

$$G_x - \frac{d}{dt} G_{x'} = y' U \ , \qquad G_y - \frac{d}{dt} G_{y'} = -x' U \ ,$$

we get

$$\chi_a = \int_{t_5}^{t} U(\psi_t \phi_a - \phi_t \psi_a)\, dt + \Big[\mathsf{G}_{x'} \phi_a + \mathsf{G}_{y'} \psi_a\Big]_{t_5}^{t} - \mathsf{G}\Big|^{t_5} \frac{\partial t_5}{\partial a} \ .$$

The terms outside of the sign of integration reduce to

$$\mathsf{G}_{x'} \phi_a + \mathsf{G}_{y'} \psi_a \Big|^{t} \ ,$$

on account of (42).

A similar transformation applies to χ_b.

We substitute these values of χ_t, χ_a, χ_b in $\Delta(t, a, b)$ and then put $a = a_0$, $b = b_0$, which makes $t_5 = t_{50}$.

Writing for brevity

$$A = \psi_t \phi_a - \phi_t \psi_a \Big|^{a=a_0}_{b=b_0} \ , \qquad B = \psi_t \phi_b - \phi_t \psi_b \Big|^{a=a_0}_{b=b_0} \ ,$$

$$M = \int_{t_5}^{t} UA\, dt \ , \qquad N = \int_{t_5}^{t} UB\, dt \ ,$$

we obtain for the Jacobian the expression[2]

$$\Delta(t, a_0, b_0) = MB - NA \ . \tag{45}$$

It is now easy to establish the relation (44); for if we substitute in one of the differential equations (8) for x, y the functions $\phi(t, a, b)$, $\psi(t, a, b)$, differentiate with respect to a and then put $a = a_0$, $b = b_0$, we get

$$\Psi(A) + \lambda_1 U = 0 \ ;$$

similarly:

$$\Psi(B) + \lambda_2 U = 0 \ .$$

[1] Compare equation (18) of §25.

[2] KNESER, *Mathematische Annalen*, Vol. LV (1902), p. 95.

Hence if we set

$$\bar{u} = \lambda_2 A - \lambda_1 B \ ,$$

$$\bar{v} = -\frac{\lambda_1}{\lambda_1^2+\lambda_2^2} A - \frac{\lambda_2}{\lambda_1^2+\lambda_2^2} B \ ,$$

$\bar{u}$ and $\bar{v}$ satisfy the same differential equations as the functions u, v introduced under b). Moreover, $\bar{u}$ and $\bar{v}$ vanish for $t = t_{50}$, since, on account of (42),

$$A(t_{50}) = 0 \ , \qquad B(t_{50}) = 0 \ .$$

Hence it follows that

$$\bar{u} = cu(t, t_{50}) \ , \qquad \bar{v} = v(t, t_{50}) + c'u(t, t_{50})$$

where c and c' are constants. Taking now $D(t, t_{50})$ in the form corresponding to (37) we obtain immediately the relation (44).

d) Mayer's law of reciprocity for isoperimetric problems: The problem: To maximize or minimize the integral J while the integral K remains constant, and the "reciprocal problem": To maximize or minimize K while J remains constant, lead to *the same totality of extremals.*[1]

For, if we distinguish the quantities referring to the second problem by a stroke and make the substitution

$$\bar{\lambda} = \frac{1}{\lambda} \ , \tag{46}$$

we have

$$\bar{H} = \frac{H}{\lambda} \ ,$$

which shows that the differential equations for the two problems become identical by the substitution $\bar{\lambda} = 1/\lambda$.

Now suppose that in both problems the given end-points are the same and that, moreover, the values prescribed in the two problems for the second integral are such that one and the same extremal $\mathfrak{E}_0$, for which $\lambda_0 \neq 0$, satisfies the

[1] This remark had already been made by EULER; see STÄCKEL, *Abhandlungen aus der Variationsrechnung*, I, p. 102.

initial conditions for both problems. Then *the equivalence of the two problems still holds for the second variation.*

For since

$$\bar{H}_1 = \frac{H_1}{\lambda} \ , \tag{47}$$

$\bar{H}_1$ has a permanent sign so long as H_1 has, and *vice versa.* The sign is the same if λ is positive, the opposite if λ is negative.

Further, the conjugate to the point 0 is the same in both problems:

$$\bar{t}_0' = t_0' \ . \tag{48}$$

For the system Σ of extremals through the point 0 is the same in both problems.

Besides

$$\bar{U} = T \ ;$$

hence since the extremal $\mathfrak{E}_0$ satisfies the differential equation

$$T + \lambda_0 U = 0 \ ,$$

we have, along $\mathfrak{E}_0$:

$$\bar{U} = -\lambda_0 U \ ,$$

and therefore, according to (45),

$$\bar{\Delta}(t, a_0, b_0) = -\lambda_0 \Delta(t, a_0, b_0) \ , \tag{49}$$

which proves our statement.

This result is due to A. MAYER, and has been called by him the *law of reciprocity for isoperimetric problems.*[1]

e) EXAMPLE XIII (see p. 210): From the expression (12) for H_1 it follows that λ must be negative in case of a maximum. Equation (13) shows, then, that the vector from any point of the curve to the center must be to the left[2] of the positive tangent. Of the two arcs which satisfy the differential equation and the initial condi-

[1] *Mathematische Annalen*, Vol. XIII (1878), p. 60; compare also KNESER, *Lehrbuch*, pp. 131 and 136.

[2] If, as we always suppose, the positive y-axis lies to the left of the positive x-axis.

tions only the one above the x-axis satisfies this condition. This arc may be represented in the form

$$\left.\begin{array}{l} x = \alpha_0 - \lambda_0 \cos t \\ y = \beta_0 - \lambda_0 \sin t \end{array}\right\} t_0 \leqq t \leqq t_1 < t_0 + 2\pi \ . \tag{50}$$

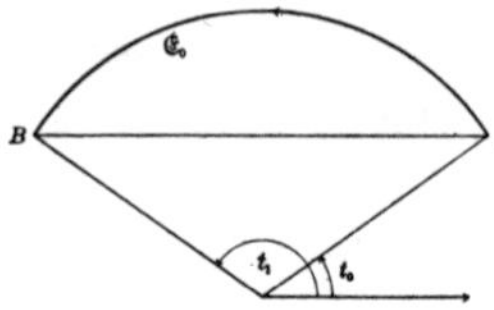
FIG. 45

Hence we obtain

$$\theta_1(t) = -\lambda_0 \cos t \ ,$$
$$\theta_2(t) = -\lambda_0 \sin t \ ,$$
$$\theta_3(t) = \lambda_0 \ .$$

Again,

$$U = \frac{x'y'' - x''y'}{(\sqrt{x'^2 + y'^2})^3} \ ,$$

which is equal to $-1/\lambda_0$ along $\mathfrak{C}_0$, according to (13). This leads to the following expression for $D(t, t_0)$:

$$D(t, t_0) = 4\lambda^2 \sin \omega (\sin \omega - \omega \cos \omega) \ , \tag{51}$$

where

$$\omega = \frac{t - t_0}{2} \ .$$

Hence we easily infer that the parameter t_0' of the conjugate point is:

$$t_0' = t_0 + 2\pi \ . \tag{52}$$

The arc $\mathfrak{C}_0$ satisfies, therefore, the condition

$$t_1 < t_0' \ .$$

On the other hand, in the problem to maximize the integral

$$\int_{t_0}^{t_1} \left[\tfrac{1}{2}(xy' - x'y) + \lambda_0 \sqrt{x'^2 + y'^2}\right] dt \ ,$$

without an isoperimetric condition, the conjugate point t_0'' is determined by the equation

$$\Theta(t, t_0) \equiv -\lambda_0^2 \sin (t - t_0) = 0 \ ,$$

whence[1]

$$t_0'' = t_0 + \pi \ ,$$

[1] The same result follows from the geometrical interpretation of Jacobi's criterion: The extremals through A are circles of radius λ_0; their envelope is a circle about A of radius $2\lambda_0$, which is touched by each circle $\mathfrak{C}$ through A at the point diametrically opposite to A on $\mathfrak{C}$.

so that, in accordance with the general theory,

$$t_0' > t_0'' .$$

f) EXAMPLE XIV (see p. 211): We have here

$$H_1 = \frac{y+\lambda}{(\sqrt{x'^2+y'^2})^3} ; \tag{53}$$

hence for a minimum it is necessary that

$$y+\lambda > 0 .$$

Of the two solutions (14) of the differential equation (I) which satisfy the initial conditions, only the one in which the upper sign is taken in the expression for $y+\lambda$, fulfils this condition.

For this solution we obtain

$$\theta_1(t) = \beta_0 \sinh t , \quad \theta_2(t) = \beta_0(t \sinh t - \cosh t) , \quad \theta_3(t) = \beta_0 ,$$

$$U = \frac{x'y''-x''y'}{(\sqrt{x'^2+y'^2})^3} = \frac{1}{\beta_0 \cosh^2 t} .$$

Hence follows

$$\int_{t_0}^{t} U\theta_1\, dt = \left[-\frac{1}{\cosh t} \right]_{t_0}^{t} ,$$

$$\int_{t_0}^{t} U\theta_2\, dt = \left[-\frac{t}{\cosh t} \right]_{t_0}^{t} ,$$

$$\int_{t_0}^{t} U\theta_3\, dt = \left[\tanh t \right]_{t_0}^{t_1} ,$$

and the expression for $D(t, t_0)$ reduces to[1]

$$D(t, t_0) = \beta_0^2 \big(2 \cosh (t-t_0) - 2 - (t-t_0) \sinh (t-t_0)\big) , \tag{54}$$

or, if we put

$$t - t_0 = 2\omega ,$$

$$D(t, t_0) = 4\beta_0^2 \sinh \omega (\sinh \omega - \omega \cosh \omega) . \tag{54a}$$

The function $\sinh \omega$ is positive for every positive ω, and the function

$$\phi(\omega) = \sinh \omega - \omega \cosh \omega$$

is negative for every positive ω, since $\phi(\omega)=0$ and

$$\phi'(\omega) = -\omega \sinh \omega .$$

[1] First given by A. MAYER, *Mathematische Annalen*, Vol. XIII (1878), p. 67.

Hence *there exists no conjugate point*, and the third necessary condition is always satisfied.

The same result is even more easily obtained by using KNESER's method:[1]

If we let the point 5 coincide[2] with the point 0 and choose for the two parameters a, b the quantities

$$a = t_5 , \qquad b = \beta ,$$

the system of extremals through the point 0 is represented by the equations

$$x - x_0 = b(t - a) , \qquad y - y_0 = b(\cosh t - \cosh a) , \qquad (55)$$

Hence we obtain

$$\chi(t, a, b) = \int_a^t \sqrt{x'^2 + y'^2}\, dt = b(\sinh t - \sinh a) , \qquad (56)$$

and therefore

$$\Delta(t, a, b) = b^2 \Big[2\cosh(t-a) - 2 - (t-a)\sinh(t-a) \Big] ,$$

which for $a = a_0 (= t_0)$, $b = b_0 (= \beta_0)$ reduces to the expression (54) for $D(t, t_0)$.

§42. SUFFICIENT CONDITIONS

The argumentation of §28 applies, with slight modifications,[3] to the present problem, and leads to a fourth necessary condition for a minimum:

[1] Compare KNESER, *Lehrbuch*, p. 143. [2] Compare the introductory lines of §41, c).

[3] These modifications are:

1. The variations ξ, η must now satisfy the isoperimetric condition:

$$\overline{K}_{04} + \widetilde{K}_{42} = K_{02} ,$$

in addition to the conditions stated in §28, a). To obtain such variations, let

$$p_i, q_i \ (i = 1, 2, 3)$$

be arbitrary functions of t of class C' satisfying the conditions:

$$p_i(t_0) = 0 , \quad q_i(t_0) = 0 , \quad p_1(t_2) = 0 , \quad q_1(t_2) = 0 ,$$
$$p_2(t_2)\, q_3(t_2) - p_3(t_2)\, q_2(t_2) \neq 0 , \quad N_1 \neq 0 ,$$

N_i having the same signification as in §39, a). Then the functions

$$\xi = \epsilon_1 p_1 + \epsilon_2 p_2 + \epsilon_3 p_3 , \qquad \eta = \epsilon_1 q_1 + \epsilon_2 q_2 + \epsilon_3 q_3$$

will satisfy all the required conditions if $\epsilon_1, \epsilon_2, \epsilon_3$ are determined by the equations

$$\xi(t_2) = \xi_2 , \quad \eta(t_2) = \eta_2 , \quad \overline{K}_{04} - K_{02} = -\widetilde{K}_{42} ,$$

which is always possible under the above assumptions concerning p_i, q_i.

2. ΔJ has to be replaced by $\Delta J + \lambda_0 \Delta K$.

If we denote by $\mathbf{E}(x, y; p, q; \tilde{p}, \tilde{q} | \lambda)$ the function derived from $H = F + \lambda G$ exactly in the same manner in which the **E**-function for the unconditioned problem is derived from the function F (see equation (48) of §28), then *the fourth necessary condition for a minimum consists in the inequality*[1]

$$\mathbf{E}(x, y; p, q; \tilde{p}, \tilde{q} | \lambda_0) \geqq 0 \tag{IV}$$

which must be fulfilled along[2] the arc $\mathfrak{E}_0$ for every direction $\tilde{p}, \tilde{q}$.

The question arises now whether the four conditions (I)–(IV) are *sufficient* for a minimum.

a) Weierstrass's construction: Let

$$\bar{\mathfrak{C}}: \qquad \bar{x} = \bar{\phi}(s), \qquad \bar{y} = \bar{\psi}(s), \qquad s_0 \leqq s \leqq s_1, \tag{57}$$

be any curve of class C', different from $\mathfrak{E}_0$, joining the points 0 and 1, lying in the region[3] $\mathfrak{R}$ and satisfying likewise the isoperimetric condition

$$\bar{K}_{01} = l;$$

for s we take for simplicity the arc of the curve $\bar{\mathfrak{C}}$.

We propose to express the difference

$$\Delta J = \bar{J}_{01} - J_{01}$$

in terms of the **E**-function.

For this purpose we take a point 5 on the continuation of the arc $\mathfrak{E}_0$ beyond 0, but not on $\bar{\mathfrak{C}}$, and consider with KNESER[4] the doubly infinite system Σ of extremals through the point 5:

$$\mathfrak{E}: \qquad x = \phi(t, a, b), \qquad y = \psi(t, a, b) \tag{58}$$

introduced in §41, *c*), the arc $\mathfrak{E}_0$ being given by

$$x = \phi(t, a_0, b_0), \qquad y = \psi(t, a_0, b_0), \qquad t_0 \leqq t \leqq t_1.$$

1 WEIERSTRASS, *Lectures*, 1879.

2 In the same sense as in §28, *a*).

3 Compare §24, *b*) and §39.

4 WEIERSTRASS considers instead the set of extremals through 0. Compare p. 240, footnote 1.

We shall say that for the curve $\bar{\mathfrak{C}}$ *Weierstrass's construction is possible*[1] if the point 5 can be so chosen that the following conditions are fulfilled:

A) Through every point 2 of the curve $\bar{\mathfrak{C}}$ there passes a uniquely defined extremal $\mathfrak{E}_2$ of the system Σ:

$$\mathfrak{E}_2: \qquad x = \phi(t, a_2, b_2), \qquad y = \psi(t, a_2, b_2), \tag{59}$$

lying wholly in the region $\mathfrak{R}$ and such that the integral K taken along $\mathfrak{E}_2$ from 5 to 2 has the same value as when taken from 5 to 0 along $\mathfrak{E}_0$ and then from 0 to 2 along $\bar{\mathfrak{C}}$:

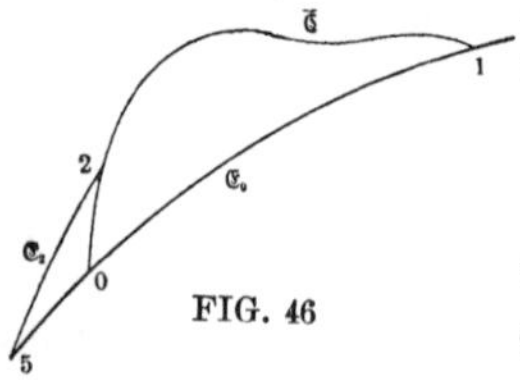

FIG. 46

$$K_{52} = K_{50} + \bar{K}_{02}; \tag{60}$$

and when 2 coincides with 0 or 1, the extremal $\mathfrak{E}_2$ coincides with $\mathfrak{E}_0$.

This means analytically: There exists a system of three single-valued functions

$$t = t(s), \qquad a = a(s), \qquad b = b(s)$$

such that

$$\begin{aligned} \phi\big(t(s), a(s), b(s)\big) &= \bar{\phi}(s), \\ \psi\big(t(s), a(s), b(s)\big) &= \bar{\psi}(s), \\ \chi\big(t(s), a(s), b(s)\big) &= \bar{\chi}(s) + K_{50} \end{aligned} \tag{61}$$

where $\chi(t, a, b)$ has the same signification as in equation (43), and

$$\bar{\chi}(s) = \int_{s_0}^{s} G\big(\bar{\phi}(s), \bar{\psi}(s), \bar{\phi}'(s), \bar{\psi}'(s)\big)\, ds.$$

Moreover:

$$\begin{aligned} t(s_0) &= t_0, & a(s_0) &= a_0, & b(s_0) &= b_0, \\ t(s_1) &= t_1, & a(s_1) &= a_0, & b(s_1) &= b_0. \end{aligned} \tag{62}$$

B) The three functions $t(s)$, $a(s)$, $b(s)$ are of class C' in $(s_0 s_1)$.

[1] Compare KNESER, *Lehrbuch*, p. 133.

C) If s_2 be any value of s of the interval $(s_0 s_1)$ and we denote:[1]

$$t_2 = t(s_2)\ , \qquad a_2 = a(s_2)\ , \qquad b_2 = b(s_2)\ , \qquad t_{52} = t_5(a_2, b_2)\ ,$$

then the functions

$$\phi,\ \psi,\ \phi_t,\ \psi_t,\ \phi_{tt},\ \psi_{tt}$$

and their first partial derivatives with respect to a and b are continuous in the domain

$$t_{52} \leqq t \leqq t_2\ , \qquad |a - a_2| \leqq d_2\ , \qquad |b - b_2| \leqq d_2\ ,$$

d_2 being a sufficiently small positive quantity, and moreover the function[2] $\lambda(a, b)$ is continuous at (a_2, b_2).

These conditions admit of the following geometrical interpretation:[3]

We adjoin to the two equations (58) the equation

$$z = \chi(t, a, b). \tag{58a}$$

Interpreting then x, y, z as rectangular co-ordinates in space, the equations (58) and (58a) represent a curve in space, $\mathfrak{C}'$, whose projection upon the x, y-plane is the extremal $\mathfrak{C}$, and whose z-co-ordinate indicates at every point t the value of the integral K taken along $\mathfrak{C}$ from the point 5 to the point t.

We thus obtain, corresponding to the system Σ, a doubly infinite system Σ' of curves in space, all passing through the point 5:

$$x = x_5\ , \qquad y = y_5\ , \qquad z = 0\ .$$

The particular curve $\mathfrak{C}_0'$ adjoined to the curve $\mathfrak{C}_0$ passes, besides, through the two points $0'$ and $1'$:

$$0':\quad x = x_0\ , \qquad y = y_0\ , \qquad z = z_0 = K_{50}\ ,$$
$$1':\quad x = x_1\ , \qquad y = y_1\ , \qquad z = z_1 = K_{50} + l\ .$$

In like manner we adjoin to the curve $\bar{\mathfrak{C}}$ a curve in space,

[1] For the notation see §41, c). [2] Compare §41, c).

[3] WEIERSTRASS, *Lectures*, 1879; compare also KNESER, *Lehrbuch*, p. 140.

$\bar{\mathfrak{C}}'$, by combining with the two equations (57) the third equation

$$z = \bar{\chi}(s) + K_{50} . \tag{57a}$$

The curve $\bar{\mathfrak{C}}'$ passes likewise through the points $0'$ and $1'$.

The above assumptions A) and B) may then be couched in geometrical language as follows:

Through every point $2'$ of the curve $\bar{\mathfrak{C}}'$ there passes a uniquely defined curve of the set Σ'; it changes continuously as the point $2'$ describes the curve $\bar{\mathfrak{C}}'$ from $0'$ to $1'$ and coincides with $\mathfrak{E}_0'$ when $2'$ coincides with $0'$ or $1'$.

Under the assumption that Weierstrass's construction is possible for the curve $\bar{\mathfrak{C}}$, we consider as in §20, *b*) and §28, *d*) the integral J taken from 5 to an arbitrary point $2(s = s_2)$ of $\bar{\mathfrak{C}}$ along the uniquely defined extremal $\mathfrak{E}_2$, and from 2 to 1 along $\bar{\mathfrak{C}}$, and denote its value regarded as a function of s_2 by $S(s_2)$:

$$S(s_2) = J_{52} + \bar{J}_{21} .$$

Then as in §20, *b*)

$$\Delta J = -\left[S(s_1) - S(s_0)\right] .$$

The integral K taken along the same path has the constant value $l + K_{50}$:

$$K_{50} + l = K_{52} + \bar{K}_{21} ,$$

since $\bar{K}_{01} = \bar{K}_{02} + \bar{K}_{21} = l$ and $K_{52} = \bar{K}_{02} + K_{50}$. Hence it follows that we may write

$$\frac{dS(s_2)}{ds_2} = \left(\frac{dJ_{52}}{ds_2} + \lambda_2 \frac{dK_{52}}{ds_2}\right) + \left(\frac{d\bar{J}_{21}}{ds_2} + \lambda_2 \frac{d\bar{K}_{21}}{ds_2}\right) . \tag{63}$$

Proceeding now as in §28, *d*) and remembering that the extremal $\mathfrak{E}_2$ satisfies the differential equations

$$H_x - \frac{d}{dt} H_{x'} = 0 , \qquad H_y - \frac{d}{dt} H_{y'} = 0 ,$$

where

$$H = F + \lambda_2 G ,$$

we obtain the result

$$\frac{dS(s_2)}{ds_2} = -\mathbf{E}(\bar{x}_2, \bar{y}_2; p_2, q_2; \bar{p}_2, \bar{q}_2 | \lambda_2) ,$$

the direction-cosines p_2, q_2 and $\bar{p}_2$, $\bar{q}_2$ referring to the curves $\mathfrak{C}_2$ and $\bar{\mathfrak{C}}$ respectively.

The result can again easily be extended to curves $\bar{\mathfrak{C}}$ having a finite number of corners.

Thus we finally reach the result[1] that *whenever Weierstrass's construction is possible for the curve* $\bar{\mathfrak{C}}$, *Weierstrass's theorem also holds:*

$$\Delta J = \int_{s_0}^{s_1} \mathbf{E}(\bar{x}_2, \bar{y}_2; p_2, q_2; \bar{p}_2, \bar{q}_2 | \lambda_2)\, ds_2 . \tag{64}$$

b) Hence we infer that $\Delta J \geqq 0$ whenever

$$\mathbf{E}(\bar{x}_2, \bar{y}_2; p_2, q_2; \bar{p}_2, \bar{q}_2 | \lambda_2) \geqq 0 \qquad \text{throughout} \quad (s_0 s_1) .$$

If, moreover, the **E**-*function vanishes only when* $\bar{p}_2 = p_2$, $\bar{q}_2 = q_2$, *and if besides*

$$\Delta(t_2, a_2, b_2) \neq 0 \qquad \textit{along} \quad \bar{\mathfrak{C}} ,$$

ΔJ *cannot be zero, and therefore*

$$\Delta J > 0 .$$

Proof:[2] If we differentiate equations (61) with respect to s, we obtain

$$\phi_t \frac{dt}{ds} + \phi_a \frac{da}{ds} + \phi_b \frac{db}{ds} = \bar{\phi}' ,$$

$$\psi_t \frac{dt}{ds} + \psi_a \frac{da}{ds} + \psi_b \frac{db}{ds} = \bar{\psi}' ,$$

$$\chi_t \frac{dt}{ds} + \chi_a \frac{da}{ds} + \chi_b \frac{db}{ds} = \bar{\chi}' .$$

Now if $p_2 = \bar{p}_2$, $q_2 = \bar{q}_2$, we have at the point 2:

$$\bar{\phi}' = k\phi_t , \qquad \bar{\psi}' = k\psi_t , \qquad (k > 0)$$

and therefore, since[3]

[1] WEIERSTRASS, *Lectures*, 1879; compare KNESER, *Lehrbuch*, p. 134.

[2] Due to KNESER, *Lehrbuch*, p. 134.

[3] Compare §41, *c*).

$$\bar{\chi}' = \bar{G}\ , \qquad \chi_t = G\ ,$$

also[1]

$$\bar{\chi}' = k\chi_t\ ,$$

on account of the homogeneity of G.

Substituting these values in the above equations, we see that either

$$\Delta(t_2, a_2, b_2) = 0\ ,$$

or else

$$\frac{da}{ds} = 0\ , \qquad \frac{db}{ds} = 0\ .$$

Hence if

$$\Delta(t_2, a_2, b_2) \neq 0 \qquad \text{along}^2\ \bar{\mathfrak{C}}\ ,$$

a_2 and b_2 must be constant along $\bar{\mathfrak{C}}$, and, on account of (62), their constant values must be

$$a(s) = a_0\ , \qquad b(s) = b_0\ ,$$

that is: $\bar{\mathfrak{C}}$ is identical with the extremal $\mathfrak{E}_0$, which is in contradiction to our assumption that $\bar{\mathfrak{C}}$ shall not coincide with $\mathfrak{E}_0$. Hence the statement is proved.

c) In many examples the above theorem is sufficient to establish the existence of an extremum.

EXAMPLE XIII (see p. 229): The system Σ is the *totality of circles through the point 5:*

$$\begin{aligned} x - x_5 &= b(\cos t - \cos a)\ , \\ y - y_5 &= b(\sin t - \sin a)\ , \end{aligned} \tag{65}$$

the parameters being $a = t_5$, $b = -\lambda$.

The ordinate z erected at the point t of the circle (a, b) is the length of the arc of this circle from the point 5 $(t = a)$ to the point t:

$$z = |b(t - a)|\ . \tag{66}$$

The system Σ' of curves in space is therefore a system of *helices*.

Through every point (x, y, z) for which

$$z > \sqrt{(x - x_5)^2 + (y - y_5)^2} > 0\ , \tag{67}$$

[1] This means geometrically: If $\mathfrak{E}_2$ touches $\bar{\mathfrak{C}}$, then also $\mathfrak{E}_2'$ touches $\mathfrak{C}'$.

[2] The result remains true if $\Delta(t_2, a_2, b_2) = 0$ at a finite number of points.

there passes one and but one curve of the system Σ' for which

$$a < t < a + 2\pi\,, \qquad b > 0\,. \tag{68}$$

Moreover the inverse functions t, a, b of x, y, z thus defined are regular[1] in the vicinity of every point (x_2, y_2, z_2) satisfying the inequality (67), and take, at the points (x_0, y_0, z_0) and (x_1, y_1, z_1) the values t_0, a_0, b_0 and t_1, a_0, b_0 respectively.[2]

Now we join the two points 0 and 1 by an ordinary curve $\bar{\mathfrak{C}}$, whose length has the given value l and which does not pass through 5.

Then for every point 2 of $\bar{\mathfrak{C}}$ the sum of the lengths of the arc 50 of the circle $\mathfrak{C}_0$ and of the arc 02 of $\bar{\mathfrak{C}}$ is greater than—never equal to—the distance between the two points 5 and 2, which in its turn is greater than zero, since $\bar{\mathfrak{C}}$ does not pass through 5, *i. e.*, the condition

[1] Proof: On setting

$$\frac{t+a}{2} = \gamma\,, \qquad \frac{t-a}{2} = \omega$$

the equations for the determination of t, a, b become

$$\begin{aligned} x - x_5 &= -2b \sin\gamma \sin\omega\,, \\ y - y_5 &= 2b \cos\gamma \sin\omega\,, \\ z &= 2b\omega\,. \end{aligned} \tag{69}$$

Hence if we put

$$\sqrt{(x - x_5)^2 + (y - y_5)^2} = u\,,$$

and suppose

$$0 < \omega < \pi\,, \qquad \text{we get} \qquad u = 2b \sin\omega\,,$$

and therefore we obtain for the determination of ω and γ the equations:

$$\frac{\sin\omega}{\omega} = v, \qquad y - y_5 - i(x - x_5) = ue^{i\gamma}\,, \tag{70}$$

where $v = u/z$. Since, according to equation (67), $0 < v < 1$, the transcendental equation for ω has one and but one solution in the interval: $0 < \omega < \pi$.

Moreover if $0 < v_2 < 1$ be any particular value of v, this solution ω is regular in the vicinity of $v = v_2$, since the derivative of the function $\sin\omega/\omega$ is $\neq 0$ for $0 < \omega < \pi$.

Similarly the equation for γ has a unique solution in the interval $0 \leqq \gamma < 2\pi$, which is a regular function of x, y in the vicinity of every point (x_2, y_2) different from (x_5, y_5).

The values of ω and γ being found, the quantities t, a, b are obtained immediately. They satisfy the inequalities (68) and are regular functions of x, y, z in the domain (67).

[2] For, of the two arcs of circles of the system Σ which pass through the point (x, y) and have the given length z, the one is described in the positive sense (so that the center is to the left) if we start from the point 5, the other in the negative sense. For the former the inequalities (68) are fulfilled, for the latter, they are not.

On the other hand the arcs 50 and 51 of $\mathfrak{C}_0$ are, according to §41, *e*), described in the positive sense, and are therefore contained in the above system of uniquely defined solutions.

$$z_2 > \sqrt{(x_2 - x_5)^2 + (y_2 - y_5)^2} > 0$$

is fulfilled.[1]

Hence it follows that Weierstrass's construction is possible for the curve $\bar{\mathfrak{C}}$.

Further we find easily that

$$\mathbf{E}(x_2, y_2; p_2, q_2; \bar{p}_2, \bar{q}_2 \mid \lambda_2) = \lambda_2(1 - \cos \alpha_2) , \tag{71}$$

where α_2 is the angle between the positive tangents to the two curves $\mathfrak{C}_2$ and $\bar{\mathfrak{C}}$ at the point 2.

λ_2 is negative in $(s_0 s_1)$ (since it is equal to $-b_2$), and α_2 cannot vanish identically in $(s_0 s_1)$.

For, according to (51),

$$\Delta(t_2, a_2, b_2) = 4\lambda_2^2 \sin \omega_2(\sin \omega_2 - \omega_2 \cos \omega_2) ,$$

and therefore

$$\Delta(t_2, a_2, b_2) \neq 0 \quad \text{in} \quad (s_0 s_1) ,$$

since $0 < \omega_2 < \pi$.

Hence it follows that

$$\Delta J < 0 ,$$

and thus we reach the result that *the arc of circle $\mathfrak{C}_0$ furnishes a greater value for the area J than any other ordinary curve of the same length which can be drawn between the two points 0 and 1.*

The same reasoning, slightly modified,[2] leads to the theorem

[1] If we had taken, instead of the system of extremals through 5, the system through 0, the above inequality would be true only with certain exceptions which would require a special discussion. Compare p. 233, footnote 4.

[2] The curve $\bar{\mathfrak{C}}$ is now closed; accordingly the points 0 and 1 coincide. If we let also the point 5 coincide with 0 and consider two points 3 and 4 of $\bar{\mathfrak{C}}$ for which $s_0 < s_3 < s_4 < s_1$, we obtain by the same reasoning as above

$$S(s_4) - S(s_3) = -\int_{s_3}^{s_4} \lambda_2(1 - \cos \alpha_2)\, ds_2 .$$

Now let s_3 and s_4 approach s_0 and s_1 respectively, then we get

$$J_{01} - \bar{J}_{01} = -\int_{s_0}^{s_1} \lambda_2(1 - \cos \alpha_2)\, ds_2 ,$$

J_{01} being the area of a circle of the given perimeter l. Hence

$$\bar{J}_{01} < J_{01} .$$

The previous method is not applicable when the curve $\bar{\mathfrak{C}}$ begins at the point 0 with a segment of a straight line, because then the inequality (67) is not satisfied for the point 3. In this case, take the point 3 beyond the end-point 6 of this rectilinear segment and let 3 approach 6. Then $S(s_3)$ approaches again $\bar{J}_{01}$ with the same result as before.

that *among all closed curves of given length the circle includes the maximum area.*

EXAMPLE XIV (see p. 231): Any admissible curve $\bar{\mathfrak{C}}$ being given, we choose the point 5 so that for every point 2 of $\bar{\mathfrak{C}}$

$$x_2 > x_5 \ .$$

Then through every point 2′ of the space curve $\bar{\mathfrak{C}}'$ one and but one curve of the system[1] Σ':

$$\begin{aligned} x - x_5 &= b\,(t - a) \ , \\ y - y_5 &= b\,(\cosh t - \cosh a) \ , \qquad (72) \\ z &= b\,(\sinh t - \sinh a) \ , \end{aligned}$$

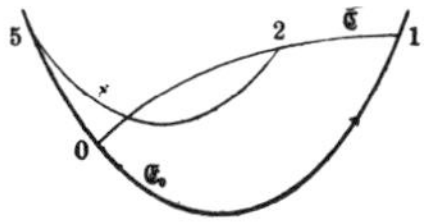

FIG. 47

can be drawn for which

$$t > a \ , \qquad b > 0 \ .$$

This follows from the determination of constants given in §39, *d*). At the same time it is easily seen, in the same manner as in the preceding example, that all the conditions for Weierstrass's construction are fulfilled.

Further we find

$$\mathbf{E}\,(x_2, y_2;\ p_2, q_2;\ \bar{p}_2, \bar{q}_2 \mid \lambda_2) = (y_2 + \lambda_2)\,(1 - \cos \alpha_2) \ , \quad (73)$$

where α_2 has the same signification as in (71). But, according to §41, *f*),

$$y_2 + \lambda_2 = b_2 \cosh t_2 > 0 \ ,$$

since $b_2 > 0$, and α_2 cannot vanish identically along $\bar{\mathfrak{C}}$ since

$$\Delta\,(t_2, a_2, b_2) \neq 0$$

along $\bar{\mathfrak{C}}$. Hence we infer that

$$\bar{J}_{01} > J_{01} \ , \qquad i.\ e.,$$

the catenary $\mathfrak{C}_0$ has its center of gravity lower than any other ordinary curve of equal length which can be drawn between the two points 0 and 1.

d) "*Field*" *about the arc* $\mathfrak{C}_0'$: Returning now to the general case, we meet with a peculiar difficulty which has

[1] Compare equations (55) and (56).

no analogue in the unconditioned problem. Suppose that for the arc $\mathfrak{C}_0$, which we assume to be free from multiple points, the conditions

$$H_1 > 0 \tag{II'}$$

and

$$t_1 < t_0' \tag{III'}$$

are fulfilled.

Does it follow, then, that the arc $\mathfrak{C}_0$ can be surrounded by a neighborhood (ρ) such that for every admissible curve $\bar{\mathfrak{C}}$ which lies wholly in this neighborhood, Weierstrass's construction is possible?

In the unconditioned problem and under the analogous assumptions, this question could be answered in the affirmative;[1] *for the isoperimetric problem the question has not yet been answered.*

Only the following milder statement can be proved:

If conditions (II') and (III') are fulfilled, a neighborhood[2] (ρ') of the space curve $\mathfrak{C}_0'$ adjoined to the arc $\mathfrak{C}_0$ can be assigned such that *Weierstrass's construction is possible for every admissible curve $\bar{\mathfrak{C}}$ whose corresponding space curve lies wholly in the neighborhood (ρ') of $\mathfrak{C}_0'$.*

The proof proceeds by the following steps:

1. If conditions (II') and (III') are fulfilled, we can take the point 5 so near to 0 that for the system of extremals through the point 5 not only the conditions enumerated in §41, *c*) are satisfied, but, besides, the following:[3]

$$\Delta(t, a_0, b_0) \neq 0 \qquad \text{for} \quad t_0 \leqq t \leqq t_1 \,. \tag{74}$$

[1] Compare §28, *d*) and §34.

[2] We understand by the neighborhood (ρ') of the arc $\mathfrak{C}_0'$ the portion of space swept out by a sphere of radius ρ' whose center describes the arc $\mathfrak{C}_0'$.

[3] For the proof remember (44), and notice that the condition for a permanent sign of $\delta^2 J$ may also be written

$$D(t_1, t) \neq 0 \qquad \text{for} \qquad t_0 \leqq t < t_1 ,$$

(compare §41, *a*)). The statement follows then by a slight modification of the analogous proof given by C. JORDAN, *Cours d'Analyse*, Vol. III, No. 393.

2. By an extension of the method of §34 we can now prove the existence of a "field" $\mathfrak{S}_k'$ about the arc $\mathfrak{E}'$:

If $\mathfrak{D}_k$ denotes the domain

$$t_0 - \epsilon \leqq t \leqq t_1 + \epsilon\,, \qquad |a - a_0| \leqq k\,, \qquad |b - b_0| \leqq k\,,$$

and $\mathfrak{S}_k'$ the image of $\mathfrak{D}_k$ in the x, y, z-space defined by the transformation

$$x = \phi(t, a, b)\,, \qquad y = \psi(t, a, b)\,, \qquad z = \chi(t, a, b)\,,$$

then the two positive quantities k and ϵ can be taken so small that the correspondence between $\mathfrak{D}_k$ and $\mathfrak{S}_k'$ is a one-to-one correspondence, and that at the same time

$$\Delta(t, a, b) \neq 0 \tag{75}$$

in $\mathfrak{D}_k$.

The single-valued functions t, a, b of x, y, z thus defined are of class C' in $\mathfrak{S}_k'$, and a neighborhood (ρ') of the arc $\mathfrak{E}_0'$ can be inscribed in $\mathfrak{S}_k'$.

It follows now easily that for every admissible curve $\overline{\mathfrak{C}}$ *whose adjoined space curve lies wholly in the "field"* $\mathfrak{S}_k'$, Weierstrass's construction is possible.

e) Sufficient conditions for a semi-strong minimum: Suppose now that in addition to the conditions (II′) and (III′) the inequality

$$\mathbf{E}(x, y;\, p, q;\, \tilde{p}, \tilde{q}\,|\,\lambda_0) > 0 \tag{IV'}$$

holds along the arc $\mathfrak{E}_0$ for every direction $\tilde{p}$, $\tilde{q}$ except $\tilde{p} = p$, $\tilde{q} = q$.

Then it follows from continuity considerations that we can take k so small that

$$\mathbf{E}(\bar{x}_2, \bar{y}_2;\, p_2, q_2;\, \bar{p}_2, \bar{q}_2\,|\,\lambda_2) > 0$$

along every admissible curve $\overline{\mathfrak{C}}$ satisfying the above additional condition, except at the points where $\bar{p}_2 = p_2$, $\bar{q}_2 = q_2$, at which $\mathbf{E}$ vanishes.

From Weierstrass's theorem and the inequality (75) it follows now that for every such curve $\bar{\mathfrak{C}}$

$$\Delta J > 0 .$$

Hence, if we modify our original definition of a minimum and say: "The arc $\mathfrak{C}_0$ furnishes a *semi-strong minimum* for the integral J if there exists a neighborhood (ρ') of the adjoined arc $\mathfrak{C}_0'$ such that $\Delta J \geqq 0$ for every admissible curve $\bar{\mathfrak{C}}$ whose adjoined space curve $\bar{\mathfrak{C}}'$ lies wholly in this neighborhood (ρ')," we can enunciate the

Theorem:[1] *The extremal $\mathfrak{C}_0$ (which we suppose free from multiple points) furnishes a semi-strong minimum for the integral J with the isoperimetric condition $K = l$, if the conditions (II'), (III'), (IV') are fulfilled.*

It must, however, be admitted that the restriction which we impose in the "semi-strong" minimum upon the variations of the arc $\mathfrak{C}_0$, is rather artificial and alters completely the character of the original problem.[2]

[1] WEIERSTRASS, *Lectures*, 1882; compare KNESER, *Lehrbuch*, §§36 and 38.

Mayer's law of reciprocity extends to the sufficient conditions for a semi-strong extremum, since, in the notation of §41, *d*), $\bar{\mathbf{E}} = 1/\lambda \mathbf{E}$. Compare KNESER, *Lehrbuch*, § 36.

[2] As a matter of fact the preceding theorem does not contain a solution of the isoperimetric problem originally proposed, but a solution of the following problem, which is usually (but unjustly) considered as equivalent to the isoperimetric problem, viz.:

Among all curves in space which pass through the two points

$$x = x_0, \ y = y_0, \ z = 0 \qquad \text{and} \qquad x = x_1, \ y = y_1, \ z = l$$

and satisfy the differential equation

$$\frac{dz}{dt} = G(x, y, x', y'),$$

to determine the one which maximizes or minimizes the integral

$$J = \int_{t_0}^{t_1} F(x, y, x', y')\, dt .$$

CHAPTER VII

HILBERT'S EXISTENCE THEOREM

§43. INTRODUCTORY REMARKS

IF a function $f(x)$ is defined for an interval (ab), it has in this interval a lower (upper) limit, finite or infinite, which may or may not be reached. If, however, the function is continuous in (ab), then the lower (upper) limit is always finite and is always reached at some point of the interval: the function has a minimum (maximum) in the interval.

Similarly, if the integral

$$J = \int_{t_0}^{t_1} F(x, y, x', y')\, dt$$

is defined for a certain manifoldness $\mathfrak{M}$ of curves, we can, in general, not say *a priori* whether the values of the integral have a minimum or maximum. But the question arises whether it is not perhaps possible to impose such restrictions either upon the function F or upon the manifoldness $\mathfrak{M}$ (or upon both), that the existence of an extremum can be ascertained *a priori*.

In a communication to the "Deutsche Mathematiker-Vereinigung" (*Jahresberichte*, Vol. VIII (1899), p. 184), HILBERT has answered this question in the affirmative. He makes the following general statement:

"Eine jede Aufgabe der Variationsrechnung besitzt eine Lösung, sobald hinsichtlich der Natur der gegebenen Grenzbedingungen geeignete Annahmen erfüllt sind und nötigenfalls der Begriff der Lösung eine sinngemässe Erweiterung erfährt," and illustrates the gist of his method by the example of the shortest line upon a surface and by Dirichlet's problem. In a subsequent[3] course of lectures (Göttingen, summer, 1900) he gave the details of his method for the shortest line on a surface, and some indications[1] concerning its extension to the problem of minimizing the integral

$$J = \int_{x_0}^{x_1} F(x, y, y')\, dx \ .$$

We propose to apply, in this last chapter, Hilbert's method to the problem of minimizing the integral[2]

$$J = \int_{t_0}^{t_1} F(x, y, x', y')\, dt \ ,$$

with fixed end-points, under the following assumptions, where $\mathfrak{R}$ denotes, as before, a region of the x, y-plane, and $\mathfrak{R}_0$ a finite closed region contained in the interior of $\mathfrak{R}$:

A) The function $F(x, y, x', y')$ *is of class* C''' and satisfies the *homogeneity condition*

$$F(x, y, kx', ky') = kF(x, y, x', y') \ , \qquad k > 0$$

throughout the domain

$$\mathfrak{T}: \qquad (x, y) \quad \text{in } \mathfrak{R} \ , \qquad x'^2 + y'^2 \neq 0 \ .$$

B) *The function* $F(x, y, \cos\gamma, \sin\gamma)$ *is positive* throughout the domain

$$\mathfrak{T}_0: \qquad (x, y) \quad \text{in } \mathfrak{R}_0 \ , \qquad 0 \leqq \gamma \leqq 2\pi \ .$$

C) *The function* $F_1(x, y, \cos\gamma, \sin\gamma)$ *is positive* throughout the domain $\mathfrak{T}_0$.

[1] In his thesis, *Eine neue Methode in der Variationsrechnung* (Göttingen, 1901), §§5-14, NOBLE has discussed the details of the proof for this case. But his conclusions do not possess the degree of rigor which is indispensable in an investigation of this kind. In particular, the reasoning in §§9, 10 and 13 is open to serious objections.

[2] For the special case where F is of the form $f(x, y)\sqrt{x'^2 + y'^2}$, LEBESGUE has given a rigorous existence proof by an elegant modification of Hilbert's method in a recent paper, "Integrale, longueur, aire," *Annali di Matematica* (3), Vol. VII (1902), pp. 342-359. LEBESGUE applies Hilbert's method also to the more difficult case of a double integral of the form

$$\int\int \sqrt{EG - F^2}\, du dv \ .$$

[3] HILBERT, *Abh.*, Vol. III, p. 15.

D) The region $\mathfrak{R}_0$ is *convex* (*i. e.*, the straight line joining any two points of $\mathfrak{R}_0$ lies entirely in the region $\mathfrak{R}_0$) and contains the two given points which we denote[1] with HILBERT by A^0 and A^1.

Under these assumptions we propose to prove

1. That for every *rectifiable* curve $\mathfrak{L}$ in the region $\mathfrak{R}_0$ the *generalized integral* $J^*_{\mathfrak{L}}$ (according to WEIERSTRASS's definition) *has a determinate finite value.*

2. That there always exists, in the region $\mathfrak{R}_0$, at least one rectifiable curve $\mathfrak{L}_0$, joining the two given points A^0 and A^1, which furnishes for the generalized integral $J^*_{\mathfrak{L}}$ an *absolute minimum with respect to the totality of all rectifiable curves which can be drawn in* $\mathfrak{R}_0$ *from* A^0 *to* A^1.

3. That this minimizing curve $\mathfrak{L}_0$ is either a single arc of an *extremal of class* C'', or else is made up of a finite number or of a numerable[3] infinitude of such arcs separated by points or segments of the boundary of the region $\mathfrak{R}_0$.

§44. THEOREMS CONCERNING THE GENERALIZED INTEGRAL $J^*_{\mathfrak{L}}$

In §31 we have considered Weierstrass's extension of the meaning of the definite integral

$$J = \int_{t_0}^{t_1} F(x, y, x', y')\, dt$$

to curves having no tangent.

Another definition of the generalized integral has been given by HILBERT[2] in his lectures. This definition, while

[1] The advantage of this notation will appear in §45.

[2] HILBERT's own definition is as follows (see NOBLE, *loc. cit.*, p. 18). Let Π_1 be a partition of the arc AB of a continuous curve into segments. Consider the totality of all analytic curves which can be drawn from A to B and which have at least one point in common with each of the segments. Let J_1 denote the lower limit of the values of the integral J taken along these curves. Next, let Π_2 be a new partition derived from Π_1 by subdivision, J_2 the corresponding lower limit, and so on. Then HILBERT defines the upper limit of the quantities: $J_1, J_2, J_3, \cdots, J_n, \cdots$ if it be finite, as the value of the definite integral J taken along the arc AB.

[3] Compare E. I A, p. 186.

leading to the same value for the generalized integral as Weierstrass's definition, is better adapted to our present purpose, especially in the simplified form which has been given to it by OSGOOD.[1]

a) Hilbert-Osgood's definition of the generalized integral: We shall use the following notation: P' and P'' being any two points of the region $\mathfrak{R}_0$, we denote by $\mathfrak{M}(P'P'')$ the totality of all ordinary curves which can be drawn in the region $\mathfrak{R}_0$ from P' to P'', and by $i(P'P'')$ the lower limit of the values which the integral

$$J = \int F(x, y, x', y')\, dt$$

takes along the various curves of $\mathfrak{M}(P'P'')$.

This lower limit is always positive. For, according to A) and B), the function $F(x, y. \cos\gamma, \sin\gamma)$ has a positive minimum value m in the closed domain $\mathfrak{T}_0$. Hence, if $\mathfrak{C}$ be any curve of $\mathfrak{M}(P'P'')$, we obtain, by taking the arc as independent variable on the curve $\mathfrak{C}$,

$$0 < m\,|P'P''| \leqq ml \leqq J_{\mathfrak{C}}(P'P'') \,, \tag{1}$$

where l denotes the length of the curve $\mathfrak{C}$ and $|P'P''|$ the distance between the two points P', P''. Hence it follows that

$$0 < m\,|P'P''| \leqq i(P'P'') \,. \tag{2}$$

After these preliminaries, let

$$\mathfrak{L}: \qquad x = \phi(t) \,, \qquad y = \psi(t) \,, \qquad t_0 \leqq t \leqq t_1$$

be a continuous curve lying wholly in the region $\mathfrak{R}_0$. If the functions $\phi(t)$, $\psi(t)$ are not differentiable, the integral J taken along $\mathfrak{L}$ has no meaning. In order to give it a meaning also in this case, we consider any partition Π of the interval $(t_0 t_1)$

[1] OSGOOD, *Transactions of the American Mathematical Society*, Vol. II (1901), p. 294, footnote.

$$\Pi: \qquad t_0 < \tau_1 < \tau_2 \cdots < \tau_{n-1} < t_1 ,$$

and denote by

$$A = P_0, P_1, P_2, \cdots, P_{n-1}, B = P_n$$

the corresponding points of the curve $\mathfrak{L}$.

Then we form the sum

$$S_\Pi = \sum_{\nu=0}^{n-1} i(P_\nu P_{\nu+1}) .$$

The upper limit of the values of S_Π for all possible partitions Π we define as the value of the integral J taken along the curve $\mathfrak{L}$ from A to B, and we denote it by $J_{\mathfrak{L}}^{**}(AB)$, or simply $J_{\mathfrak{L}}^{**}$.

It is easily seen that S_Π may also be defined[1] as the lower limit of the values of the integral J taken along all ordinary curves which can be drawn in $\mathfrak{R}_0$ from A to B and which pass in succession through the points $P_1, P_2, \cdots, P_{n-1}$.

Hence it follows that it is always possible to select a sequence $\{\mathfrak{C}_\nu\}$ of ordinary curves joining A and B, lying in $\mathfrak{R}_0$, and such that

$$\underset{\nu=\infty}{L}\, J_{\mathfrak{C}_\nu} = J_{\mathfrak{L}}^{**} .$$

The above definition of the generalized integral is a direct generalization of PEANO's[2] definition of the length of a curve. For, in the particular case

$$F = \sqrt{x'^2 + y'^2} ,$$

the sum S_Π reduces to the length of the rectilinear polygon with the vertices $A, P_1, P_2, \cdots, P_{n-1}, B$.

We must next investigate under what conditions the generalized integral $J_{\mathfrak{L}}^{**}$ is finite, and show that for ordinary

[1] This is the form which OSGOOD gives to Hilbert's definition; see the reference on p. 248, footnote 1.

[2] PEANO, *Applicazioni geometriche del Calcolo Infinitesimale*, p. 161.

curves the generalized integral is identical with the ordinary definite integral.

b) Conditions for the finiteness of the generalized integral: The function $F(x, y, \cos\gamma, \sin\gamma)$ has a finite maximum value M in the domain $\mathfrak{T}_0$. Hence it follows that for every curve $\mathfrak{C}$ of $\mathfrak{M}(P'P'')$

$$i(P'P'') \leqq J_{\mathfrak{C}}(P'P'') \leqq Ml \ , \tag{2a}$$

l denoting again the length of the curve $\mathfrak{C}$. We may choose for the curve $\mathfrak{C}$ the straight line $P'P''$, since, according to assumption D), the line $P'P''$ lies wholly in the region $\mathfrak{R}_0$. Then we obtain the further inequality

$$i(P'P'') \leqq M|P'P''| \ . \tag{3}$$

From (2) and (3) follows at once

$$m\sum_{\nu=0}^{n-1}|P_\nu P_{\nu+1}| \leqq S_{\mathrm{II}} \leqq M\sum_{\nu=0}^{n-1}|P_\nu P_{\nu+1}| \ . \tag{4}$$

But the upper limit of the sum

$$\sum_{\nu=0}^{n-1}|P_\nu P_{\nu+1}|$$

is, according to Peano's definition, the length of the curve $\mathfrak{L}$. Hence we obtain the

Lemma: In order that the generalized integral $J_{\mathfrak{L}}^{**}$ may be finite, it is necessary and sufficient that the curve $\mathfrak{L}$ shall have a finite length (in Peano's sense).

We confine ourselves, therefore, in the sequel to continuous curves $\mathfrak{L}$ having a finite length ("rectifiable curves" in JORDAN's terminology).[1] From (4) it follows further that

$$m|AB| \leqq J_{\mathfrak{L}}^{**}(AB) \leqq ML \ , \tag{5}$$

where L denotes the length of the curve $\mathfrak{L}$.

[1] J. I, No. 110.

c) Properties of the generalized integral: From the two characteristic properties of the lower limit it follows readily that for any three points P, P', P'' of $\mathfrak{R}_0$ the inequality holds:

$$i(PP') + i(P'P'') \geqq i(PP'') \ . \tag{6}$$

Hence it follows that if Π_1 denotes a partition derived from Π by *subdivision* of the intervals of Π, then

$$S_{\Pi_1} \geqq S_{\Pi} \ .$$

Hence we easily infer that we get the same upper limit $J^{**}_{\mathfrak{L}}$ for the values of S_{Π} if we confine ourselves to those partitions Π for which

$$\tau_{\nu+1} - \tau_{\nu} < \delta \ ,$$
$$(\nu = 0, 1, 2, \cdots, n-1; \ \tau_0 = t_0, \tau_n = t_1) \ ,$$

δ being an arbitrary positive quantity.

Following now step by step the same reasoning which JORDAN uses in his discussion of the length of a curve, we can easily establish the following properties of the generalized integral, always under the assumption that the curve $\mathfrak{L}$ is rectifiable:

1. The generalized integral $J^{**}_{\mathfrak{L}}(AB)$ is at the same time the limit which the sum S_{Π} approaches as all the differences $\tau_{\nu+1} - \tau_{\nu}$ approach zero.[1]

Combining this result with the inequality (4) we obtain the new inequality

$$mL \leqq J^{**}_{\mathfrak{L}}(AB) \ . \tag{7}$$

2. If P be a point on the curve $\mathfrak{L}$ between A and B, dividing the arc $\mathfrak{L}$ into the two arcs $\mathfrak{L}_1$ and $\mathfrak{L}_2$, then also the integrals $J^{**}_{\mathfrak{L}_1}(AP)$ and $J^{**}_{\mathfrak{L}_2}(PB)$ are finite, and[2]

$$J^{**}_{\mathfrak{L}}(AB) = J^{**}_{\mathfrak{L}_1}(AP) + J^{**}_{\mathfrak{L}_2}(PB) \ . \tag{8}$$

3. The generalized integral $J^{**}_{\mathfrak{L}}(AP)$ is a continuous[3]

[1] Compare J. I, No. 107. [2] Compare J. I, No. 108. [3] Apply (8) and (5).

function of the parameter t of the point P and increases continually as P describes the arc AB from A to B.

d) Comparison with Weierstrass's definition of the generalized integral: If P' and P'' are two points of $\mathfrak{R}_0$ whose distance from each other is less than the quantity ρ_0 defined at the end of §28, *e*), P' and P'' can be joined by an extremal $\mathfrak{E}$ of class C' which furnishes for the integral J a smaller value than any other ordinary curve which can be drawn in the region $\mathfrak{R}_0$ from P' to P''. If the extremal $\mathfrak{E}$ itself lies entirely in the region $\mathfrak{R}_0$, the value which it furnishes for the integral J is equal to $i(P'P'')$; if $\mathfrak{E}$ lies partly outside of $\mathfrak{R}_0$, this value is equal to or less than $i(P'P'')$.

Now consider any partition Π for which

$$\tau_{\nu+1} - \tau_\nu < \delta, \ (\nu = 0, 1, \cdots, n-1) ,$$

δ being chosen so small that $|P'P''| < \rho_0$ for any two points P', P'' of $\mathfrak{L}$ whose parameters t', t'' satisfy the inequality $|t' - t''| < \delta$. Then we can inscribe in the curve $\mathfrak{L}$ *a polygon of minimizing extremals* with the vertices

$$A, P_1, P_2, \cdots, P_{n-1}, B .$$

As in §31, *d*), let U_Π denote the value of the integral J taken along this polygon of extremals.

If the curve $\mathfrak{L}$ lies entirely in the interior of $\mathfrak{R}_0$, δ can be taken so small that the polygon lies in the region $\mathfrak{R}_0$, and therefore

$$U_\Pi = S_\Pi .$$

Hence $J^{**}_{\mathfrak{L}}$ may in this case also be defined as the limit of U_Π.

If $\mathfrak{L}$ has points in common with the boundary of $\mathfrak{R}_0$, U_Π may be less than S_Π.

Nevertheless, also in this case the limit of U_Π for $\Delta\Delta\tau = 0$ is $J^{**}_{\mathfrak{L}}$.

In order to prove this statement we consider, along with

the two sums S_Π and U_Π, the sum V_Π defined in §31, *c*), *i. e.*, the value of the integral J taken along the rectilinear polygon $AP_1P_2\cdots P_{n-1}B$. Since the region $\mathfrak{R}_0$ is *convex*, this polygon lies entirely in $\mathfrak{R}_0$, and therefore we have the double inequality

$$U_\Pi \leqq S_\Pi \leqq V_\Pi \ . \tag{9}$$

From the first part of this inequality it follows that U_Π has a finite upper limit $\leqq J_{\mathfrak{L}}^{**}$. This upper limit is at the same time the limit which U_Π approaches for $L\Delta\tau = 0$, as can be inferred[1] from the fact (proved in §31, *e*)) that if Π' be a partition derived from Π by subdivision, then $U_{\Pi'} \geqq U_\Pi$. Hence it follows, according to §31, *c*) and *d*), that V_Π approaches the same limit as U_Π; therefore we obtain, on account of (9), and remembering the equations (77) and (80) of §31:

$$J_{\mathfrak{L}}^{**} = J_{\mathfrak{L}}^{*} \ , \tag{10}$$

i. e., we have the result **due to Osgood that *Hilbert-Osgood's definition leads for the generalized integral to the same value as Weierstrass's definition.***

Hence it follows, according to §31, *b*), that *for an "ordinary" curve the generalized integral coincides with the ordinary definite integral.*

§45. HILBERT'S CONSTRUCTION

We are now prepared to apply HILBERT'S method to the integral[2] $J_{\mathfrak{L}}^{*}$.

Accordingly we consider the totality of all *rectifiable* curves $\mathfrak{L}$ which can be drawn in the region $\mathfrak{R}_0$ from the point A^0 to the point A^1. The corresponding values of the integral $J_{\mathfrak{L}}^{*}$ have a positive[3] lower limit. We propose to

[1] Compare J. I, No. 107.

[2] On account of (10) we may use the symbol $J_{\mathfrak{L}}^{*}$ instead of $J_{\mathfrak{L}}^{**}$.

[3] According to (5).

prove that under the assumptions A)–D) enumerated in §43, *there exists at least one rectifiable curve* $\mathfrak{L}_0$ *drawn in* $\mathfrak{R}_0$ *from* A^0 *to* A^1 *for which the integral* $J^*_{\mathfrak{L}}$ *actually reaches its lower limit.*

a) Construction of the point $A^{\frac{1}{2}}$*:* We consider the totality of ordinary curves $\mathfrak{M}(A^0A^1)$ which can be drawn in the region $\mathfrak{R}_0$ from A^0 to A^1, and denote the lower limit $i(A^0A^1)$ of the corresponding values of the integral J by K·

$$i(A^0A^1) = K \; .$$

We can then select[1] an infinite sequence of curves

$$\mathfrak{C}_1, \mathfrak{C}_2, \cdots, \mathfrak{C}_\nu, \cdots,$$

belonging to $\mathfrak{M}(A^0A^1)$ such that the corresponding sequence of values of the integral J, which we denote by

$$J_1, J_2, \cdots, J_\nu, \cdots,$$

approaches K as limit:

$$\underset{\nu=\infty}{L} J_\nu = K \; .$$

On the curve $\mathfrak{C}_\nu$ there exists[2] one and but one point $A^{\frac{1}{2}}_\nu$ such that

$$J_{\mathfrak{C}_\nu}(A^0 A^{\frac{1}{2}}_\nu) = \tfrac{1}{2} J_\nu \; .$$

These points $A^{\frac{1}{2}}_\nu$ are infinite in number;[3] they lie in the finite[4]

[1] Compare JORDAN's definition of "point limite,' *loc. cit.*, No. 20, and an analogous remark in E. II A, p. 14.

[2] Since F is positive along $\mathfrak{C}_\nu$ the integral J taken along the curve $\mathfrak{C}_\nu$ from A^0 to a variable point P, increases continually as P describes the curve $\mathfrak{C}_\nu$ from A^0 to A^1; hence it passes through every value between 0 and J_ν once and but once.

[3] They need not all be distinct; the conclusion holds even if there are only a finite number of distinct points among them. For in this case an infinitude of the points $A^{\frac{1}{2}}_\nu$ must coincide with at least one of the distinct points; this point has then the properties of the point $A^{\frac{1}{2}}$.

[4] The existence of the accumulation point $A^{\frac{1}{2}}$ can also be proved *without making use of the finiteness of* $\mathfrak{R}_0$. From (1) it follows that

$$|A^0 A^{\frac{1}{2}}_\nu| \leqq \frac{J_\nu}{2m} \; .$$

Hence if we select $G > J_\nu \; (\nu = 1, 2, 3, \cdots)$, which is always possible since $\underset{\nu=\infty}{L} J_\nu$ is finite, the points $A^{\frac{1}{2}}_\nu$ lie in the interior of the circle $(A^0, G/2m)$, and therefore have an accumulation point.

closed region $\mathfrak{R}_0$; hence there must exist at least one point $A^{\frac{1}{2}}$ in $\mathfrak{R}_0$ such that every vicinity of $A^{\frac{1}{2}}$ contains an infinitude of the points $A^{\frac{1}{2}}_{\nu}$. Moreover, we can select a subsequence $\{\mathfrak{C}_{\nu_k}\}$ of the sequence $\{\mathfrak{C}_{\nu}\}$ such that

$$\underset{k=\infty}{L} A^{\frac{1}{2}}_{\nu_k} = A^{\frac{1}{2}} .$$

b) Hilbert's lemma concerning the point $A^{\frac{1}{2}}$*:* We consider next the totality of curves

$$\mathfrak{M}(A^0 A^{\frac{1}{2}}) .$$

Then the fundamental lemma holds that the lower limit of the corresponding values of the integral J is $\frac{1}{2}K$:

$$i(A^0 A^{\frac{1}{2}}) = \tfrac{1}{2} i(A^0 A^1) = \tfrac{1}{2} K . \tag{11}$$

Proof: We denote by $\mathfrak{C}'_{\nu_k}$ the curve made up of the arc $A^0 A^{\frac{1}{2}}_{\nu_k}$ of the curve $\mathfrak{C}_{\nu_k}$ and of the straight line $A^{\frac{1}{2}}_{\nu_k} A^{\frac{1}{2}}$; the latter lies entirely in $\mathfrak{R}_0$ since $\mathfrak{R}_0$ is convex.

According to (2a) the integral J taken along the straight line $A^{\frac{1}{2}}_{\nu_k} A^{\frac{1}{2}}$ is at most equal to $M|A^{\frac{1}{2}}_{\nu_k} A^{\frac{1}{2}}|$. Therefore

$$\underset{k=\infty}{L} J_{\mathfrak{C}'_{\nu_k}}(A^0 A^{\frac{1}{2}}) = \tfrac{1}{2} K$$

since

$$\underset{k=\infty}{L} \tfrac{1}{2} J_{\nu_k} = \tfrac{1}{2} K \qquad \text{and} \quad \underset{k=\infty}{L} | A^{\frac{1}{2}}_{\nu_k} A^{\frac{1}{2}} | = 0 .$$

Hence it follows from the characteristic properties of the lower limit that

$$i(A^0 A^{\frac{1}{2}}) \leqq \tfrac{1}{2} K .$$

In the same way we prove that

$$i(A^{\frac{1}{2}} A^1) \leqq \tfrac{1}{2} K .$$

But, on the other hand, according to (6):

$$i(A^0 A^{\frac{1}{2}}) + i(A^{\frac{1}{2}} A^1) \geqq i(A^0 A^1) .$$

The three inequalities are compatible only if separately:

$$i(A^0 A^{\frac{1}{2}}) = \tfrac{1}{2} K \qquad \text{and} \quad i(A^{\frac{1}{2}} A^1) = \tfrac{1}{2} K .$$

c) The points $A^{q/2^n}$*:* Repeating the process of section *a)* with the points A^0 and $A^{\frac{1}{2}}$ we obtain a new point, $A^{\frac{1}{4}}$, lying in the region $\mathfrak{R}_0$ and having the characteristic property that

$$i(A^0A^{\frac{1}{4}}) = i(A^{\frac{1}{4}}A^{\frac{1}{2}}) = \tfrac{1}{2}\, i(A^0A^{\frac{1}{2}}) = \tfrac{1}{4}K .$$

In like manner we derive from the two points $A^{\frac{1}{2}}$ and A^1 a point, $A^{\frac{3}{4}}$, satisfying the relation

$$i(A^{\frac{1}{2}}A^{\frac{3}{4}}) = i(A^{\frac{3}{4}}A^1) = \tfrac{1}{2}\, i(A^{\frac{1}{2}}A^1) = \tfrac{1}{4}K .$$

By an indefinite repetition of this process we obtain an infinite set of points

$$\left\{ A^{\frac{q}{2^n}} \right\}, \qquad \begin{array}{l} q = 0, 1, 2, \cdots, 2^n , \\ n = 0, 1, 2, \cdots \end{array}$$

all lying in the region $\mathfrak{R}_0$ and having the characteristic property that

$$i\left(A^{\frac{q}{2^n}} A^{\frac{q+1}{2^n}}\right) = \tfrac{1}{2^n} K . \tag{12}$$

More generally

$$i(A^{\tau'}A^{\tau''}) = (\tau'' - \tau')K , \tag{13}$$

if

$$\tau' = \frac{q'}{2^{n'}} , \qquad \tau'' = \frac{q''}{2^{n''}} ,$$

where n', n'' are integers, q', q'' odd integers, and

$$0 \leqq \tau' < \tau'' \leqq 1 .$$

For, reducing τ' and τ'' to the same denominator

$$\tau' = \frac{q}{2^n} , \qquad \tau'' = \frac{q+r}{2^n} ,$$

we obtain, according to (6) and (12),

$$i\left(A^0 A^{\frac{q}{2^n}}\right) \leqq \sum_{\mu=0}^{q-1} i\left(A^{\frac{\mu}{2^n}} A^{\frac{\mu+1}{2^n}}\right) = \frac{q}{2^n} K ,$$

$$i\left(A^{\frac{q}{2^n}} A^{\frac{q+r}{2^n}}\right) \leqq \sum_{\mu=q}^{q+r-1} i\left(A^{\frac{\mu}{2^n}} A^{\frac{\mu+1}{2^n}}\right) = \frac{r}{2^n} K ,$$

$$i\left(A^{\frac{q+r}{2^n}} A^1\right) \leqq \sum_{\mu=q+r}^{2^n-1} i\left(A^{\frac{\mu}{2^n}} A^{\frac{\mu+1}{2^n}}\right) = \frac{2^n - q - r}{2^n} K ,$$

whence
$$i\left(A^0 A^{\frac{q}{2^n}}\right) + i\left(A^{\frac{q}{2^n}} A^{\frac{q+r}{2^n}}\right) + i\left(A^{\frac{q+r}{2^n}} A^1\right) \leqq K \ .$$

But on the other hand, we have, on account of (6),
$$K = i\left(A^0 A^1\right) \leqq i\left(A^0 A^{\frac{q}{2^n}}\right) + i\left(A^{\frac{q}{2^n}} A^{\frac{q+r}{2^n}}\right) + i\left(A^{\frac{q+r}{2^n}} A^1\right) \ .$$

The two inequalities are compatible only if in each of the above three formulae the equality sign holds, which proves (13).

From (2) and (13) follows the important inequality
$$|A^{\tau'} A^{\tau''}| \leqq (\tau'' - \tau') \frac{K}{m} \ , \tag{14}$$

where $|A^{\tau'} A^{\tau''}|$ denotes again the distance between the two points $A^{\tau'}$, $A^{\tau''}$.

Let us now denote by
$$x(\tau) \ , \qquad y(\tau)$$

the rectangular co-ordinates of the point A^{τ}, τ being one of the fractions $q/2^n$ considered above. Then
$$|x(\tau') - x(\tau'')| \leqq |A^{\tau'} A^{\tau''}| \ , \qquad |y(\tau') - y(\tau'')| \leqq |A^{\tau'} A^{\tau''}| \ ,$$

and therefore on account of (14)
$$\left.\begin{aligned} |x(\tau') - x(\tau'')| &\leqq (\tau'' - \tau') \frac{K}{m} \ , \\ |y(\tau') - y(\tau'')| &\leqq (\tau'' - \tau') \frac{K}{m} \ . \end{aligned}\right\} \tag{15}$$

d) The remaining points of Hilbert's curve: The meaning of the two functions $x(t)$, $y(t)$, which so far have been defined only for values of t of the set
$$S = \left\{ \frac{q}{2^n} \right\} \ , \qquad \begin{aligned} q &= 0, 1, 2, \cdots, 2^n - 1 \ , \\ n &= 0, 1, 2, \cdots, \end{aligned}$$

can now be extended to all values of t in the interval
$$0 \leqq t \leqq 1$$

as follows:

From the inequalities (15) we infer by means of the general criterion[1] for the existence of a limit, that if the independent variable t approaches *in the set* S any particular value $t=a$ of the interval (01), then the functions $x(t)$, $y(t)$ approach determinate finite limits. In symbols, the limits[2]

$$\underset{\substack{t\,|\,S\\ t=a}}{L}\, x(t) \qquad \text{and} \qquad \underset{\substack{t\,|\,S\\ t=a}}{L}\, y(t)$$

exist and are finite.

Moreover, if a itself belongs to the set S, then

$$\underset{\substack{t\,|\,S\\ t=a}}{L}\, x(t) = x(a) \ , \qquad \underset{\substack{t\,|\,S\\ t=a}}{L}\, y(t) = y(a) \ . \tag{16}$$

If a does not belong to the set S, we define, according to Hilbert, the functions $x(t)$ and $y(t)$ for $t=a$ by the equations (16).

The two functions $x(t)$, $y(t)$ thus defined for the whole interval (01) are *continuous* and "*of limited variation.*"[3] For, the two inequalities (15), which have been proved for values $\tau'<\tau''$ *of the set* S, can easily be shown to hold for any two values $t'<t''$ of the interval (01), by considering two sequences $\{\tau_\nu'\}$ and $\{\tau_\nu''\}$ belonging to the set S and such that

$$\underset{=\infty}{L}\, \tau_\nu' = t' \ , \qquad \underset{\nu=\infty}{L}\, \tau_\nu'' = t'' \ .$$

From the inequalities (15) thus extended, it follows at once that the two functions $x(t)$, $y(t)$ are continuous and "of limited variation."

[1] Compare E. II A, p. 13.

[2] The notation according to E. H. MOORE, *Transactions of the American Mathematical Society*, Vol. I (1900), p, 500.

[3] Compare J. I, No. 67. Let $f(t)$ be finite in the interval $(t_0 t_1)$, and let

$$\Pi : \qquad t_0 < \tau_1 < \tau_2 \cdot \cdot \cdot < \tau_{n-1} < t_1$$

be a partition of this interval. If then the upper limit of the sum

$$\sum_{\nu=0}^{n-1} |f(\tau_{\nu+1}) - f(\tau_\nu)| \ , \qquad (\tau_0 = t_0 \, , \, \tau_n = t_1)$$

for all possible partitions Π is finite, $f(t)$ is said to be "of limited variation."

Hence the curve $\mathfrak{L}_0$ defined by the two equations

$$\mathfrak{L}_0: \qquad x = x(t), \qquad y = y(t), \qquad 0 \leqq t \leqq 1 \tag{17}$$

is continuous and has a finite length,[1] *i. e.*, it is a *rectifiable curve.* As t increases from 0 to 1 the point (x, y) describes the curve $\mathfrak{L}_0$ from the point A^0 to the point A^1. Moreover, the *curve* $\mathfrak{L}_0$ *lies entirely in the region* $\mathfrak{R}_0$, since $\mathfrak{R}_0$ is *closed.*

§46. PROPERTIES OF HILBERT'S CURVE

It remains now to prove that the curve $\mathfrak{L}_0$ actually minimizes the integral J^* and has the further properties stated in §43.

a) Minimizing property of Hilbert's curve: The fundamental equation (13) which has been proved for values τ', τ'' of the set S only, can easily be extended[2] to any two values $t' < t''$ of the interval (01):

$$i(A^{t'}A^{t''}) = (t'' - t')K. \tag{13a}$$

But from (13a) it follows immediately that *the generalized integral*

[1] Compare J. I, Nos 105, 110.

[2] For the proof, we introduce the same two sequences $\{\tau'_\nu\}$, $\{\tau''_\nu\}$ as above. Then we have, on account of (6),

$$i(A^{t'}A^{\tau'_\nu}) + i(A^{\tau'_\nu}A^{\tau''_\nu}) + i(A^{\tau''_\nu}A^{t''}) \geqq i(A^{t'}A^{t''}).$$

Passing to the limit $\nu = \infty$ we obtain, on account of the continuity of the functions $x(t), y(t)$,

$$\underset{\nu=\infty}{L} |A^{t'}A^{\tau'_\nu}| = 0, \qquad \underset{\nu=\infty}{L} |A^{\tau''_\nu}A^{t''}| = 0,$$

and therefore, on account of (3),

$$\underset{\nu=\infty}{L} i(A^{t'}A^{\tau'_\nu}) = 0, \qquad \underset{\nu=\infty}{L} i(A^{\tau''_\nu}A^{t''}) = 0.$$

Moreover

$$\underset{\nu=\infty}{L} i(A^{\tau'_\nu}A^{\tau''_\nu}) = (t'' - t')K,$$

on account of (13). Thus we obtain

$$i(A^{t'}A^{t''}) \leqq (t'' - t')K.$$

And by the method employed in proving (13) we finally show that the inequality sign is impossible.

$$J^*_{\mathfrak{L}_0}(A^0A^1)$$

taken along HILBERT'S *curve* $\mathfrak{L}_0$ *is finite and that its value is equal to* $i(A^0A^1)$.

For let Π be any partition of the interval (01):

$$\Pi: \qquad \tau_0 = 0 < \tau_1 < \tau_2 \cdots < \tau_{n-1} < 1 = \tau_n \ .$$

Then we obtain, according to (13a),

$$S_\Pi = \sum_{\nu=0}^{n-1} i\,(A^{\tau_\nu} A^{\tau_{\nu+1}}) = K = i\,(A^0A^1) \ .$$

Hence also the upper limit of the values of S_Π is equal to K, that is

$$J^*_{\mathfrak{L}_0}(A^0A^1) = i\,(A^0A^1) \ . \tag{18}$$

From the definition of the symbol $i(A^0A^1)$ as lower limit it follows now that if $\mathfrak{C}$ be any ordinary curve drawn in $\mathfrak{R}_0$ from A^0 to A^1, then

$$J^*_{\mathfrak{L}_0}(A^0A^1) \leqq J_{\mathfrak{C}}(A^0A^1) \ .$$

Moreover, if $\mathfrak{L}$ be any rectifiable curve drawn in $\mathfrak{R}_0$ from A^0 to A^1, and ϵ any preassigned positive quantity, we can always find, according to §44, *a*), an ordinary curve $\mathfrak{C}$ of $\mathfrak{M}(A^0A^1)$ such that

$$| J^*_{\mathfrak{L}}(A^0A^1) - J_{\mathfrak{C}}(A^0A^1) | < \epsilon \ .$$

Hence it follows that

$$J^*_{\mathfrak{L}_0}(A^0A^1) \leqq J^*_{\mathfrak{L}}(A^0A^1) \ . \tag{19}$$

This proves the theorem enunciated at the beginning of this section:

If the conditions A)–D) enumerated in §43 are fulfilled, then there always exists at least one rectifiable curve joining the two points A^0 and A^1 and lying entirely in the region $\mathfrak{R}_0$, which furnishes for the integral

$$J = \int F(x, y, x', y')\, dt \ ,$$

generalized, an absolute minimum with respect to the totality of rectifiable curves which can be drawn in $\mathfrak{R}_0$ from A^0 to A^1.

b) Analytic character of Hilbert's curve: Let T' denote the totality of those values of t in the interval (01) which furnish points of the curve $\mathfrak{L}_0$ in the interior of the region $\mathfrak{R}_0$, T'' the totality of those which furnish points of $\mathfrak{L}_0$ on the boundary of $\mathfrak{R}_0$. From the continuity of $\mathfrak{L}_0$ it follows that every point[1] t' of T' is an inner[2] point of T'. Hence an interval $(\alpha\beta)$ contained in (01) and containing t' in its interior can be determined such that all points in the interior of $(\alpha\beta)$ belong to T', whereas the end-points belong to T'' except when they coincide with the points 0 or 1. The set T' consists, therefore, of a finite or infinite number of such intervals $(\alpha\beta)$ which do not overlap. According to a theorem of CANTOR'S,[3] the totality of these intervals is numerable, so that we may denote them by

$$\{(\alpha_\nu \beta_\nu)\} \; .$$

The curve $\mathfrak{L}_0$ consists, therefore, either of a finite number or of a numerable infinitude of interior arcs separated by points of the boundary of $\mathfrak{R}_0$

We are going to prove, according to HILBERT, that each interior arc of $\mathfrak{L}_0$ is an arc of an extremal of class[4] C''.

For let $P(t)$ be a point of Hilbert's curve $\mathfrak{L}_0$ in the interior of the region $\mathfrak{R}_0$. Then according to §28, *e*) a circle (P, σ) can be constructed[5] about P such that any two points P', P'' in the interior of the circle can be joined by an extremal $\mathfrak{E}$ of class C'' which lies entirely in the region $\mathfrak{R}_0$ and which furnishes a smaller value for the integral J than any other ordinary curve which can be drawn in $\mathfrak{R}_0$ from P' to P''.

1 Except the end-points of the interval (01) in case they should belong to T'.

2 Compare J. I, No. 22.

3 *Mathematische Annalen*, Vol. XX, p. 118.

4 From our assumption C) it follows according to §6, *c*) that every arc of an extremal of class C' which lies in $\mathfrak{R}_0$, is *ipso facto* also of class C''.

5 Let d be a positive quantity, taken so small that the circle (P, e) lies in the interior of $\mathfrak{R}_0$, and let ρ_0 be defined for the region $\mathfrak{R}_0$ as in §28, *e*). Then choose for σ the smaller of the two quantities $d/3$ and $\rho_0/3$.

On account of the continuity of the functions $x(t)$, $y(t)$ there exists a vicinity $(t-\delta,\ t+\delta)$ of t such that the arc of the curve $\mathfrak{L}_0$ corresponding to the interval[1] $(t-\delta,\ t+\delta)$ lies wholly in the interior of the circle $(P,\ \sigma)$. Let $P_1(t_1)$ and $P_3(t_3)$ be two points of this arc $(t_1 < t_3)$, and denote by $\mathfrak{E}_2$ the minimizing extremal joining P_1 and P_3.

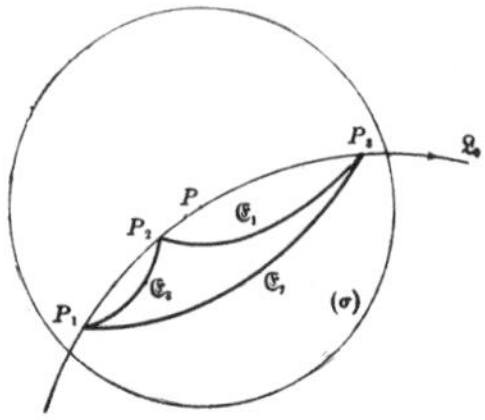

FIG. 48

We propose to prove *that the arc* P_1P_3 *of* HILBERT'S *curve* $\mathfrak{L}_0$ *is identical with the extremal* $\mathfrak{E}_2$.

Consider any point $P_2(t_2)$ of the arc P_1P_3 of $\mathfrak{L}_0$ and denote by $\mathfrak{E}_3$, $\mathfrak{E}_1$ the minimizing extremals joining P_1, P_2 and P_2, P_3 respectively. Then it follows from the minimizing properties of the extremals $\mathfrak{E}_1$, $\mathfrak{E}_2$, $\mathfrak{E}_3$ and from (13a) that

$$J_{\mathfrak{E}_3}(P_1P_2) = i(P_1P_2) = (t_2 - t_1)K\ ,$$
$$J_{\mathfrak{E}_1}(P_2P_3) = i(P_2P_3) = (t_3 - t_2)K\ ,$$
$$J_{\mathfrak{E}_2}(P_1P_3) = i(P_1P_3) = (t_3 - t_1)K\ ;$$

hence, adding:

$$J_{\mathfrak{E}_2}(P_1P_3) = J_{\mathfrak{E}_3}(P_1P_2) + J_{\mathfrak{E}_1}(P_2P_3)\ .$$

The extremal $\mathfrak{E}_2$ furnishes therefore the same value for the integral J as the curve made up of the two arcs $\mathfrak{E}_3$ and $\mathfrak{E}_1$. But this is in contradiction to the minimizing property of $\mathfrak{E}_2$ unless the compound curve $\mathfrak{E}_3$, $\mathfrak{E}_1$ coincides with $\mathfrak{E}_2$. Therefore the point P_2 must be a point of $\mathfrak{E}_2$; moreover

$$J_{\mathfrak{E}_2}(P_1P_2) = i(P_1P_2) = (t_2 - t_1)K\ .$$

Conversely, every point of the extremal $\mathfrak{E}_2$ belongs at the same time to the arc P_1P_3 of $\mathfrak{L}_0$. For, let P_4 be any point of $\mathfrak{E}_2$ between P_1 and P_3, and let

$$u = J_{\mathfrak{E}_2}(P_1P_4)\ .$$

Then

[1] Or $(0,\ \delta)$, or $(1-\delta,\ 1)$ in case P coincides with the point A^0 or A^1.

$$0 < u < J_{\mathfrak{E}_2}(P_1 P_3) = (t_3 - t_1) K \ .$$

Hence if we define t_4 by the relation

$$u = (t_4 - t_1) K \ ,$$

t_4 lies between t_1 and t_3 and is therefore the parameter of some point $\bar{P}_4$ of $\mathfrak{L}_0$ between P_1 and P_3. The point $\bar{P}_4$ belongs therefore also to $\mathfrak{E}_2$ and we have

$$J_{\mathfrak{E}_2}(P_1 \bar{P}_4) = (t_4 - t_1) K = J_{\mathfrak{E}_2}(P_1 P_4) \ .$$

Hence it follows that $\bar{P}_4$ must coincide with P_4 since F is positive along $\mathfrak{E}_2$.

Prom the relation between t_4 and the quantity u (which may be taken as the parameter on $\mathfrak{E}_2$), it follows, moreover, that the points are ordered on both arcs in the same manner, which completes the proof that the arc P_1P_3 of $\mathfrak{L}_0$ is identical with the extremal $\mathfrak{E}_2$.

Hence it follows that Hilbert's curve $\mathfrak{L}_0$ is of class C'' and satisfies Euler's differential equation in the vicinity of every interior point P, and therefore every interior arc of $\mathfrak{L}_0$ is indeed an arc of an extremal of class C''.

From the assumption B) that F is always positive it follows finally that Hilbert's curve $\mathfrak{L}_0$ can have *no multiple points*.

INDEX

INDEX

[The numbers refer to the pages, the subscripts to the footnotes.]

BASIC GEOMETRY

By G. D. BIRKHOFF and R. BEATLEY

"is in accord with the present approach to plane geometry. It offers a sound mathematical development . . . and at the same time enables the student to move rapidly into the heart of geometry."—*The Mathematics Teacher.*

"should be required reading for every teacher of Geometry."—*Mathematical Gazette.*

—3rd ed. 1959. 294 pp. 5⅜x8. 8284-0120-9.

VORLESUNGEN UEBER DIFFERENTIALGEOMETRIE, Vols. I, II

By W. BLASCHKE

TWO VOLUMES IN ONE.

Partial Contents: VOL. I. 1. Theory of Curves. 2. Extremal Curves. 3. Strips. 4. Theory of Surfaces. 5. Invariant Derivatives on a Surface. 6. Geometry on a Surface. 7. On the Theory of Surfaces in the Large. 8. Extremal Surfaces. 9. Line Geometry.

VOL. II. *Affine Differential Geometry.* 1. Plane Curves in the Small. 2. Plane Curves in the Large. 3. Space Curves. 5. Theory of Surfaces. 6. Extremal Surfaces. 7. Special Surfaces.

—3rd ed. (Vol. I); 2nd ed. (Vol. II). 1930/23-67. 589 pp. 6x9. 8284-0202-7. Two vols. in one.

KREIS UND KUGEL

By W. BLASCHKE

Isoperimetric properties of the circle and sphere, the (Brunn-Minkowski) theory of convex bodies, and differential-geometric properties (in the large) of convex bodies. A standard work.

—x + 169 pp. 5½x8½. 8284-0059-8. Cloth
8284-0115-2. Paper

INTEGRALGEOMETRIE

By W. BLASCHKE and E. KÄHLER

THREE VOLUMES IN ONE.

VORLESUNGEN UEBER INTEGRALGEOMETRIE, Vols. I and II, by *W. Blaschke.*

EINFUEHRUNG IN DIE THEORIE DER SYSTEME VON DIFFERENTIALGLEICHUNGEN, by *E. Kähler.*

—1936/37/34-49. 222 pp. 5½x8½. 8284-0064-4. Three vols in one.

FUNDAMENTAL EXISTENCE THEOREMS,
by G. A. BLISS. **See EVANS**

VORLESUNGEN UEBER FOURIERSCHE INTEGRALE

By S. BOCHNER

—1932-48. 237 pp. 5½x8½. 8284-0042-3.

HISTORY OF SLIDE RULE, *By F. CAJORI.* See BALL

INTRODUCTORY TREATISE ON LIE'S THEORY OF FINITE CONTINUOUS TRANSFORMATION GROUPS

By J. E. CAMPBELL

Partial Contents: CHAP. I. Definitions and Simple Examples of Groups. II. Elementary Illustrations of Principle of Extended Point Transformations. III. Generation of Group from Its Infinitesimal Transformations. V. Structure Constants. VI. Complete Systems of Differential Equations. VII. Diff. Eqs. Admitting Known Transf. Groups. VIII. Invariant Theory of Groups. IX. Primitive and Stationary Groups. XI. Isomorphism. XIII. Construction of Groups from Their Structure Constants . . . XIV. Pfaff's Equation . . . XV. Complete Systems of Homogeneous Functions. XVI. Contact Transformations. XVII. Geometry of C. T.'s. XVIII. Infinitesimal C. T's. XX. Differential Invariants. XXI.-XXIV. Groups of Line, of Plane, of Space. XXV. Certain Linear Groups.

—1903-66. xx + 416 pp. 5⅜x8. 8284-0183-7.

THEORY OF FUNCTIONS

By C. CARATHÉODORY

Translated by F. STEINHARDT.

Partial Contents: **Part One.** Chap. I. Algebra of Complex Numbers. II. Geometry of Complex Numbers. III. Euclidean, Spherical, and Non-Euclidean Geometry. **Part Two.** Theorems from Point Set Theory and Topology. Chap. I. Sequences and Continuous Complex Functions. II. Curves and Regions. III. Line Integrals. **Part Three.** Analytic Functions. Chap. I. Foundations. II. The Maximum-modulus principle. III. Poisson Integral and Harmonic Functions. IV. Meromorphic Functions. **Part Four.** Generation of Analytic Functions by Limiting Processes. Chap. I. Uniform Convergence. II. Normal Families of Meromorphic Functions. III. Power Series. IV. Partial Fraction Decomposition and the Calculus of Residues. **Part Five.** Special Functions. Chap. I. The Exponential Function and the Trigonometric Functions. II. Logarithmic Function. III. Bernoulli Numbers and the Gamma Function.

Vol. II.: **Part Six.** Foundations of Geometric Function Theory. Chap. I. Bounded Functions. II. Conformal Mapping. III. The Mapping of the Boundary. **Part Seven.** The Triangle Function and Picard's Theorem. Chap. I. Functions of Several Complex Variables. II. Conformal Mapping of Circular-Arc Triangles. III. The Schwarz Triangle Functions and the Modular Function. IV. Essential Singularities and Picard's Theorems.

"A book by a master . . . Carathéodory himself regarded [it] as his finest achievement . . . written from a catholic point of view."—*Bulletin of A.M.S.*

—Vol. I. 2nd ed. 1958. 310 pp. 6x9. 8284-0097-0.
—Vol. II. 2nd ed. 1960. 220 pp. 6x9. 8284-0106-3.

LECTURES ON ERGODIC THEORY

By P. R. HALMOS

CONTENTS: Introduction. Recurrence. Mean Convergence. Pointwise Convergence. Ergodicity. Mixing. Measure Algebras. Discrete Spectrum. Automorphisms of Compact Groups. Generalized Proper Values. Weak Topology. Weak Approximation. Uniform Topology. Uniform Approximation. Category. Invariant Measures. Generalized Ergodic Theorems. Unsolved Problems.

"Written in the pleasant, relaxed, and clear style usually associated with the author. The material is organized very well and painlessly presented."

—*Bulletin of the A.M.S.*

—1956-60. viii + 101 pp. 5⅜x8. 8284-0142-X.

ELEMENTS OF QUATERNIONS

By W. R. HAMILTON

Sir William Rowan Hamilton's last major work, and the second of his two treatises on quaternions.

—3rd ed. 1899/1901-68. 1,185 pp. 6x9. 8284-0219-1. Two vol. set.

RAMANUJAN: Twelve Lectures on His Life and Works

By G. H. HARDY

The book is somewhat more than an account of the mathematical work and personality of Ramanujan; it is one of the very few full-length books of "shop talk" by an important mathematician.

—1940-59. viii + 236 pp. 6x9. 8284-0136-5.

GRUNDZUEGE DER MENGENLEHRE

By F. HAUSDORFF

The original, 1914 edition of this famous work contains many topics that had to be omitted from later editions, notably, the theories of content, measure, and discussion of the Lebesgue integral. Also, general topological spaces, Euclidean spaces, special methods applicable in the Euclidean plane, the algebra of sets, partially ordered sets, etc.

—1914-49. 484 pp. 5⅜x8. 8284-0061-X.

SET THEORY

By F. HAUSDORFF

Hausdorff's classic text-book is an inspiration and a delight. The translation is from the Third (latest) German edition.

"We wish to state without qualification that this is an indispensable book for all those interested in the theory of sets and the allied branches of real variable theory."—*Bulletin of A. M. S.*

—2nd ed. 1962. 352 pp. 6x9. 8284-0119-5.

ELECTRICAL PAPERS

By O. HEAVISIDE

Heaviside's collected works are in five volumes: The two volumes of his *Electrical Papers* and the three volumes of his *Electromagnetic Theory.*

"The [forthcoming publication of my *Electromagnetic Theory*] brought the question of a reprint of the earlier papers to a crisis. For, as the later work grows out of the earlier, it seemed an absurdity to leave the earlier work behind. [It possesses] sufficient continuity of subject-matter and treatment, and even regularity of notation, to justify its presentation in the original form . . . It might be regarded . . . as an educational work for students of theoretical electricity."—*From the Preface.*

—2nd (c.) ed. 1892-1970. 1,183 pp. 5⅜x8. 8284-0235-3. Two vol. set.

ELECTROMAGNETIC THEORY

By O. HEAVISIDE

Third edition, with an Introduction by B. A. Behrend and with added notes on Heaviside's unpublished writings. A classic since its original publication in 1894-1912.

—3rd ed. 1970. 1,610 pp. 5⅜x8. 8284-0237-X. Three vol. set

VORLESUNGEN UEBER DIE THEORIE DER ALGEBRAISCHEN ZAHLEN

By E. HECKE

"An elegant and comprehensive account of the modern theory of algebraic numbers."

—*Bulletin of the A. M. S.*

—2nd ed. 1970. viii + 274 pp. 5⅜x8. 8284-0046-6.

INTEGRALGLEICHUNGEN UND GLEICHUNGEN MIT UNENDLICHVIELEN UNBEKANNTEN

By E. HELLINGER and O. TOEPLITZ

"Indispensable to anybody who desires to penetrate deeply into this subject."—*Bulletin of A.M.S.*

—1928-53. 286 pp. 5⅜x8. 8284-0089-X.

THEORIE DER ALGEBRAISCHE FUNKTIONEN EINER VARIABELN

By K. HENSEL and G. LANDSBERG

Partial Contents: PART ONE (Chaps. 1-8): Algebraic Functions on a Riemann Surface. PART TWO (Chaps. 9-13): The Field of Algebraic Functions. PART THREE (Chaps. 14-22): Algebraic Divisors and the Riemann-Roch Theorem. PART FOUR (Chaps. 23-27): Algebraic Curves. PART FIVE (Chaps. 28-31): The Classes of Algebraic Curves. PART SIX (Chaps. 32-37): Algebraic Relations among Abelian Integrals. APPENDIX: Historical Development. Geometrical Methods. Arithmetical Methods.

—1902-65. xvi + 707 pp. 6x9. 8284-0179-9.